Free Radical Chain Reactions in Organic Synthesis

BEST SYNTHETIC METHODS

Series Editors

A. R. Katritzky
University of Florida
Gainesville, Florida
USA

O. Meth-Cohn
Sunderland Polytechnic
Sunderland
UK

C. W. Rees
Imperial College of Science
and Technology
London, UK

Richard F. Heck, *Palladium Reagents in Organic Syntheses*, 1985
Alan H. Haines, *Methods for the Oxidation of Organic Compounds: Alkanes, Alkenes, Alkynes, and Arenes*, 1985
Paul N. Rylander, *Hydrogenation Methods*, 1985
Ernest W. Colvin, *Silicon Reagents in Organic Synthesis*, 1988
Andrew Pelter, Keith Smith and Herbert C. Brown, *Borane Reagents*, 1988
Basil Wakefield, *Organolithium Methods*, 1988
Alan H. Haines, *Methods for the Oxidation of Organic Compounds: Alcohols, Alcohol Derivatives, Alkyl Halides, Nitroalkanes, Alkyl Azides, Carbonyl Compounds, Hydroxyarenes and Aminoarenes*, 1988
H. G. Davies, R. H. Green, D. R. Kelly and S. M. Roberts, *Biotransformations in Preparative Organic Chemistry: The Use of Isolated Enzymes and Whole Cell Systems*, 1989
I. Ninomiya and T. Naito, *Photochemical Synthesis*, 1989
T. Shono, *Electroorganic Synthesis*, 1991
William B. Motherwell and David Crich, *Free Radical Chain Reactions in Organic Synthesis*, 1991

Free Radical Chain Reactions in Organic Synthesis

William B. Motherwell and David Crich

Departments of Chemistry,
Imperial College of Science, Technology and Medicine,
and University College,
London, UK

ACADEMIC PRESS

Harcourt Brace Jovanovich, Publishers
London   San Diego   New York
Boston   Sydney   Tokyo   Toronto

ACADEMIC PRESS LIMITED
24–28 Oval Road
London NW1 7DX

US Edition published by
ACADEMIC PRESS INC.
San Diego, CA 92101

This book is a guide providing general information concerning its subject matter; it is not a procedural manual. Synthesis of chemicals is a rapidly changing field. The readers should consult current procedural manuals for state-of-the-art instructions and applicable government safety regulations. The publisher and the authors do not accept responsibility for any misuse of this book, including its use as a procedural manual or as a source of specific instructions.

British Library Cataloguing in Publication Data

A catalogue record for this book is available from the British Library

ISBN 0-12-508760-8

Typeset by Thomson Press (India) Limited, New Delhi,
Printed in Great Britain by St Edmundsbury Press, Bury, St. Edmonds Suffolk, UK.

Contents

"And now for the Initiation Ceremony"

Dedicated with admiration, gratitude and affection to
Professor Sir Derek Barton FRS

Foreword

There is a vast and often bewildering array of synthetic methods and reagents available to organic chemists today. Many chemists have their own favoured methods, old and new, for standard transformations, and these can vary considerably from one laboratory to another. New and unfamiliar methods may well allow a particular synthetic step to be done more readily and in higher yield, but there is always some energy barrier associated with their use for the first time. Furthermore, the very wealth of possibilities creates an information-retrieval problem. How can we choose between all the alternatives, and what are their real advantages and limitations? Where can we find the precise experimental details, so often taken for granted by the experts? There is therefore a constant demand for books on synthetic methods, especially the more practical ones like *Organic Syntheses, Organic Reactions*, and *Reagents for Organic Synthesis*, which are found in most chemistry laboratories. We are convinced that there is a further need, still largely unfulfilled, for a uniform series of books, each dealing concisely with a particular topic from a *practical* point of view—a need, that is, for books full of preparations, practical hints and detailed examples, all critically assessed, and giving just the information needed to smooth our way painlessly into the unfamiliar territory. Such books would obviously be a great help to research students as well as to established organic chemists.

We have been very fortunate with the highly experienced and expert organic chemists, who, agreeing with our objective, have written the first group of volumes in this series. *Best Synthetic Methods*. We shall always be pleased to receive comments from readers and suggestions for future volumes.

A. R. K., O. M.-C., C. W. R.

Preface

In recent times the myth of free radicals as highly reactive intermediates has been exploded by the advent of rationally designed, highly efficient free radical chain sequences permitting both the high yielding interconversion of functional groups and the formation of carbon–carbon bonds under mild neutral conditions. The extent of this change in perception is such that free radical chain reactions can now be said to occupy an equal place in the armoury of the synthetic organic chemist with long standing two electron and concerted processes.

Several books and review articles dealing with various facets of this general area have appeared: the intention of this particular volume, which we hope will be useful to both the experienced practitioner and to the, as yet, un-initiated, is to provide a guide to the more practical aspects of the subject. This book does not seek to be comprehensive but rather to provide a collection of tried and tested experimental parts as well as a selection of methods of very recent vintage. We are only too well aware that free radical chain processes of the $S_{RN}1$ type or those mediated by transition metal redox couples can, and in future may well, form the material for further volumes. This is a task which we leave for others.

In the final analysis, the writing of a book is not only a labour but also a test of love. Given that the dedicatee once informed one of the authors that "those who can, do, those who can't, review," we hope that we may be forgiven. We are especially grateful to our friend Dr Tom Purcell, whose modesty forbade him to append his name to the cartoon which we commissioned from him. Finally, we would like to thank those of our colleagues, and, in particular, Dr Stephen Caddick, Dr Robyn S. Hay-Motherwell and Dr A. M. K. Pennell who have provided invaluable assistance in the painstaking task of checking this work.

WILLIAM B. MOTHERWELL
DAVID CRICH

Detailed Contents

−1−

Some Basic Concepts of Free Radical Chain Reactions

1.1 INTRODUCTION

It is a sad fact that for many generations of organic chemistry students their first exposure to a free radical chain reaction lay in the chlorination of methane. All too often the final picture which emerged was of a highly exothermic process in which selective monochlorination was difficult to achieve and hosts of free radicals were racing around the reaction vessel in a most indiscriminate way. Thankfully, this atypical characterization is finally being laid to rest.

Within the last 15 years, ever increasing numbers of synthetic organic chemists have come to appreciate and understand the true values of working with "free, but well domesticated" [1] preparative radical chain reactions. At the level of selective functional group manipulation, the relative indifference of a free radical intermediate to its immediate molecular environment and to solvent has allowed application of a given transformation over a wide range of natural product families of vastly differing structural type and polarity. Such a claim cannot be made for many ionic counterparts. Even more importantly, at the fundamental level of strategy and design in organic synthesis, key carbon–carbon bond forming reactions are now routinely considered in terms of a homolytic retrosynthetic disconnection.

The primary objective of this book is to illustrate the development and application of many of these more recently discovered radical chain reactions, and, by emphasizing and explaining many of the practical considerations which are common to all, to encourage their day to day use in the laboratory.

1.2 THE ADVANTAGES (Fig. 1.1)

With the exception of pericyclic reactions, the traditional chemistry of organic synthesis is ionic in nature and undoubtedly influenced by the early mechanistic picture featuring the two electron curly arrow. The reactive intermediates, carbocations and carbanions, by the very nature of their charge,

Neutral

Solvation is less important

Operation in polar and hindered environments is effective

FIG. 1.1. The advantages of free radicals over ions in organic synthesis.

seem particularly disposed to accept or to donate an electron pair in a reactive process. This simple picture, however, takes no account of solvation or aggregation phenomena. In reality, charged reactive intermediates tend to be much more bulky and hence very strongly influenced by the nature and polarity of the surrounding functional groups.

Moreover, the perennial problem of unwanted basicity associated with the use of carbanions as nucleophiles does not exist with neutral alkyl radicals which cannot induce epimerization of sensitive centres such as those adjacent to the carbonyl group.

By way of contrast, solvation effects [2], although they do exist, are much less important in neutral free radical reactions. Consequently, the small and penetrating radical is particularly effective in carrying out transformations in highly hindered situations or in molecules which possess many polar carbon–heteroatom bonds.

As a consequence of the above argument, a given reaction can often be successfully translated over a vast range of solvents and molecules of differing polarity, ranging from lipophilic steroids through to polar carbohydrates and peptides, with a much greater degree of confidence than in ionic systems.

1.3   THE RELATIVE "CHARACTER" OF VARIOUS RADICAL SPECIES

1.3.1   Reactivity as a function of the nature of the atom containing the unpaired electron [3]

Although free radicals are neutral entities, they do not all behave in a colourless uniform way, and their chemical reactivity is dominated, in the first instance, by the nature of the atom containing the unpaired electron. Accordingly they may then be endowed with either electrophilic or nucleophilic character and display either hard or soft behaviour.

Consider, for example, the trends within group VI of the periodic table, as illustrated by a selection of favoured and disfavoured reactions of alkoxyl, thiyl and selenyl radicals. The chemistry of alkoxyl radicals centres around their electrophilic nature, and typical reactions are hydrogen atom abstraction and β-scission (Scheme 1.1).

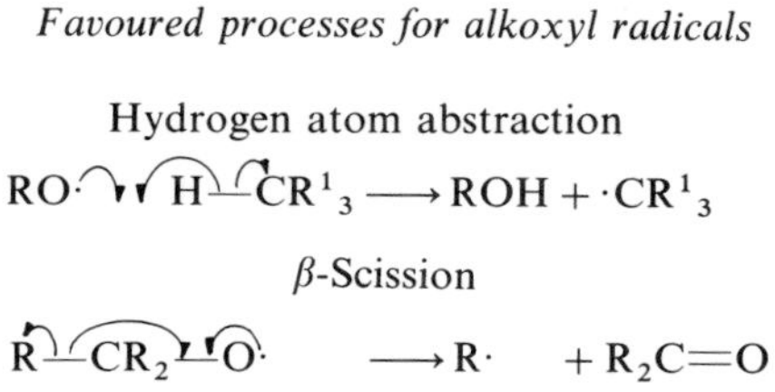

Favoured processes for alkoxyl radicals

Hydrogen atom abstraction

$$RO\cdot + H-CR^1_3 \longrightarrow ROH + \cdot CR^1_3$$

β-Scission

$$R-CR_2-O\cdot \longrightarrow R\cdot + R_2C{=}O$$

SCHEME 1.1

The thermodynamically favoured formation of the strong O—H bond at the expense of the weaker C—H bond is a powerful driving force and the intramolecular variant is a key step in the Barton nitrite ester photolysis.

In a similar fashion, β-scission leads to the creation of a carbonyl group in which the $\pi_{C=O}$ bond formed is almost as strong as a carbon–carbon σ bond. An automatic corollary is that the addition of a simple alkyl radical to a carbonyl group is not a favoured process. Neither the β-elimination of an alkoxyl radical from a neighbouring carbon centred radical nor the addition of an alkoxyl radical to a carbon–carbon double bond are synthetically useful intermolecular processes although intramolecular variants are known (Scheme 1.2). Contrastingly, thiyl radicals are larger,

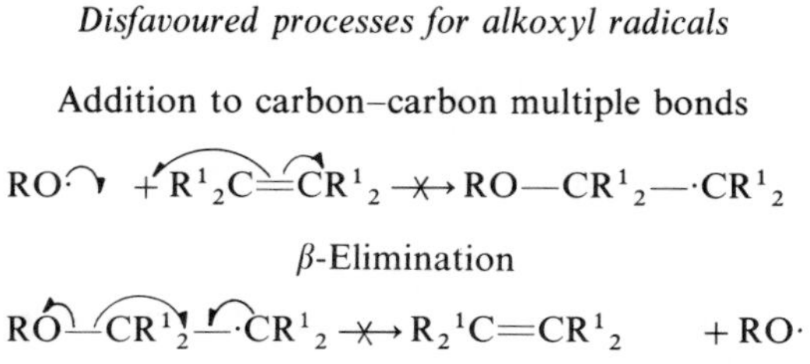

Disfavoured processes for alkoxyl radicals

Addition to carbon–carbon multiple bonds

$$RO\cdot + R^1_2C{=}CR^1_2 \nrightarrow RO-CR^1_2-\cdot CR^1_2$$

β-Elimination

$$RO-CR^1_2-\cdot CR^1_2 \nrightarrow R_2{}^1C{=}CR^1_2 + RO\cdot$$

SCHEME 1.2

softer and much less inclined to abstract hydrogen atoms. Addition to the carbon–carbon double bond is, however, a much more favoured process than for alkoxyl radicals, as is the reverse β-elimination reaction (Scheme 1.3). However, as demonstrated by the facility of photochemically induced diphenyl diselenide isomerization of olefins [4], both addition and elimination are easily reversible processes (Scheme 1.4).

Favoured processes for thiyl and selenyl radicals

Addition to carbon–carbon multiple bonds

$$RS \cdot \quad + R^1{}_2C = CR^1{}_2 \longrightarrow RSCR^1{}_2 - \dot{C}R^1{}_2$$

β-Elimination

$$RSCR^1{}_2 - \dot{C}R^1{}_2 \quad\quad \longrightarrow RS \cdot + R^1{}_2C = CR^1{}_2$$

Disfavoured processes for thiyl and selenyl radicals

Hydrogen atom abstraction

$$RS \cdot + H - CR^1{}_3 \; \nrightarrow \; RSH + \quad \cdot CR^1{}_3$$

SCHEME 1.3

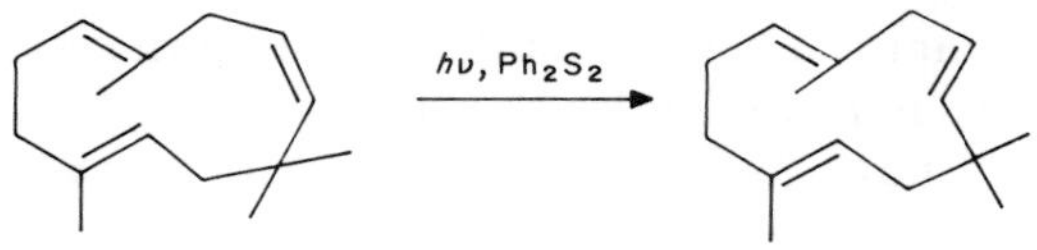

$$\xrightarrow{h\upsilon,\, Ph_2S_2}$$

SCHEME 1.4

1.3.2   Character variations within carbon centred radicals

The electrophilic or nucleophilic character is also influenced by the nature of the groupings attached to the radical centre containing the unpaired electron. The case of carbon centred radicals is, of course, particularly important since a fundamental appreciation of their differing types and behaviour lies at the very heart of organic synthesis in terms of carbon–carbon bond formation. A simple alkyl substituted carbon centred radical may be considered to be essentially nucleophilic in character because of the inductive effects of the alkyl groups, and consequently prefers to react with an electron deficient alkene such as an acrylate ester or acrylonitrile. Kinetic measurements have established, for example, that cyclohexyl radicals react some 8500 times faster with acrolein than with 1-hexene [5]. Contrastingly, the trifluoromethyl radical, as an electrophile, undergoes addition reactions most efficiently with electron rich olefins such as enamines and enol ethers. These considerations may be summarized as in Scheme 1.5.

These tendencies have been concisely rationalized using frontier molecular orbital theory [6], in which the singly occupied molecular orbital (SOMO) of the radical selects either the highest occupied molecular orbital (HOMO) or the lowest unoccupied molecular orbital (LUMO) of the alkene in the most favourable way which minimizes the energy difference between them (Fig. 1.2).

The chemoselective character of substituted alkyl radicals

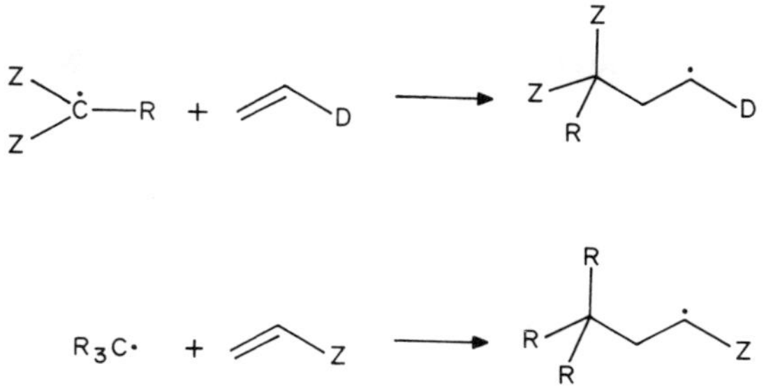

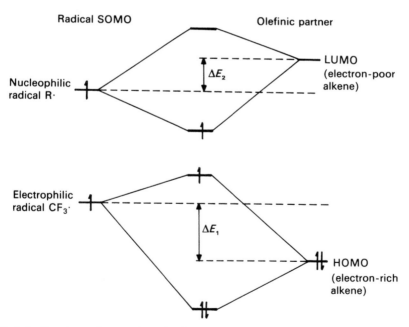

Z = electron-withdrawing group, e.g. CN, CO_2Me, SO_2Ar, etc.
D = electron-releasing group, e.g. NR_2, OR
R = alkyl group

SCHEME 1.5

FIG. 1.2. Dominant interactions in the reaction of carbon centred radicals with olefins.

Electrophilic radicals have SOMO energies which are so low that interaction with the HOMO of the electron rich alkene is dominant, whereas the effect of an electron withdrawing substituent on the olefin is to reduce the LUMO energy to such an extent that this reaction becomes favoured for nucleophilic radicals.

A simple but very important consequence of this analysis lies in the concept of radical "Umpolüng" introduced by Giese [7] (Scheme 1.6). Thus the traditional chemistry of dialkyl malonates is carbanionic in nature and Michael addition to an electron deficient olefin is the normal course of events. The carbon centred malonyl radical, however, is electrophilic and reaction with an enamine is therefore 23 times faster than with an analogous acrylic ester derivative [8].

Preferred pathways illustrating radical "Umpolüng"

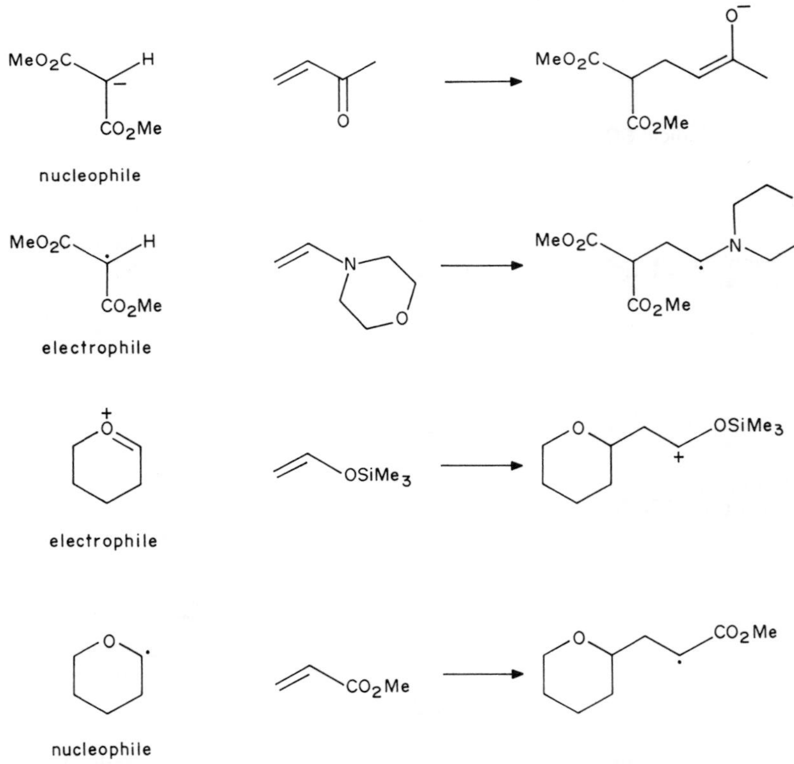

nucleophile

electrophile

electrophile

nucleophile

SCHEME 1.6

In a similar vein, replacement of the halogen atom in α-haloethers occurs readily with nucleophiles whereas the corresponding radical prefers to react with an electrophilic alkene. It is interesting to note that the relative magnitude of these effects is most clearly seen in the reaction of nucleophilic radicals with electron deficient alkenes. The discriminating reactivity of an electrophilic radical towards a range of electronically varied alkenes is much smaller, perhaps as a consequence of an inherently greater energy gap between the radical SOMO and the HOMO of an electron rich alkene, when compared with that between a radical SOMO and the LUMO of an electron deficient alkene (i.e. ΔE_1 is always inherently greater than ΔE_2 in Fig. 1.2).

1.4 DETERMINANT FACTORS IN THE PRODUCTION OF A SUCCESSFUL FREE RADICAL CHAIN REACTION

The necessary factors involved in the evolution of a free radical chain sequence are best appreciated by example. The reductive removal of a halogen atom (Section 3.1) by a tin hydride, now considered to be a standard synthetic operation, is a case in point.

$$RX \xrightarrow{\text{}^n\text{Bu}_3\text{SnH}} RH + {}^n\text{Bu}_3\text{SnX}$$

The pioneering observations of Van der Kerk [9], followed by the work of Kuivila [10], have demonstrated that this simple chain reaction is extraordinarily efficient. The formation of the strong tin–halogen bond and the rapid hydrogen atom abstraction by the alkyl radical from the weak tin–hydrogen bond (~ 20 kcal mol^{-1}) lead in a formidable thermodynamic combination to constant regeneration of the chain carrying triorganotin radical and hence to long chain lengths in the propagation sequence (Scheme 1.7).

Propagation sequence

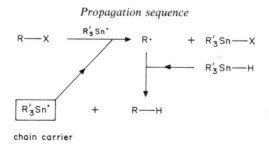

chain carrier

SCHEME 1.7

Conceptually, the most important feature of this propagation sequence is that *both steps involve the reaction of a radical with a neutral molecule*. In consequence, at any given moment in time, the concentration of organo-stannyl radicals and alkyl radicals in the reaction medium is extremely low, and the opportunity for radical–radical combination and/or dispro-portionation reactions is correspondingly more remote. This feature is particularly important since the direct combination of two radicals can occur at almost diffusion controlled rates, and often in a non-selective manner, whereas the reaction of a radical with a neutral molecule is several orders of magnitude slower. Chain reactions are inherently more elegant and controllable than their non-chain counterparts in which stoichiometric radical generation is followed by recombination with concomitant destruction of the radical character.

The fundamental importance of a thorough knowledge of relative rates in radical chain reactions cannot be overestimated; and in this respect, the earlier studies, in particular of the Ingold School [11], in providing detailed "radical clocks" have given a measure of quantitative predictability to the state of the art. In the above dehalogenation sequence the intuitive chemist can correctly anticipate the trend that alkyl iodides and bromides tend to react spontaneously, whereas chlorides may require some heating and fluorides are inert. Equally well, it is straightforward to predict that the nature of the carbon centred radical R$^{•}$ which is produced by halogen atom abstraction plays a determinant role. Thus, for alkyl radicals, the ease of generation follows the expected sequence tertiary > secondary > primary. The formation of higher energy aryl and vinyl radicals requires a higher operating temperature and the use of bromide or iodide as the halogen is now mandatory.

While knowledge of these factors, *per se*, does not influence the outcome of the reaction, quantitative estimation becomes critical on the insertion of an additional third step into the propagation sequence as illustrated by one of the prototypical intermolecular carbon–carbon bond forming reactions [12] which have been widely employed in recent years. The desired reaction, the necessary propagation sequence, and the competing reactions to avoid, are set out in Scheme 1.8. Although each of the three steps in the propagation sequence is thermodynamically favourable, an optimum yield of the desired product will only be obtained by slow addition of tri-n-butyltin hydride to the alkyl iodide in the presence of an approximately 10–100-fold excess of acrylonitrile. This *modus operandi*, which minimizes undesirable competing reactions of reduction, hydrostannylation and polymerization, was derived from quantitative knowledge of relative rates as follows:

(1) *Selection of the halogen atom.* Alkyl chlorides cannot be used as substrates since chlorine atom abstraction by the organostannyl radical

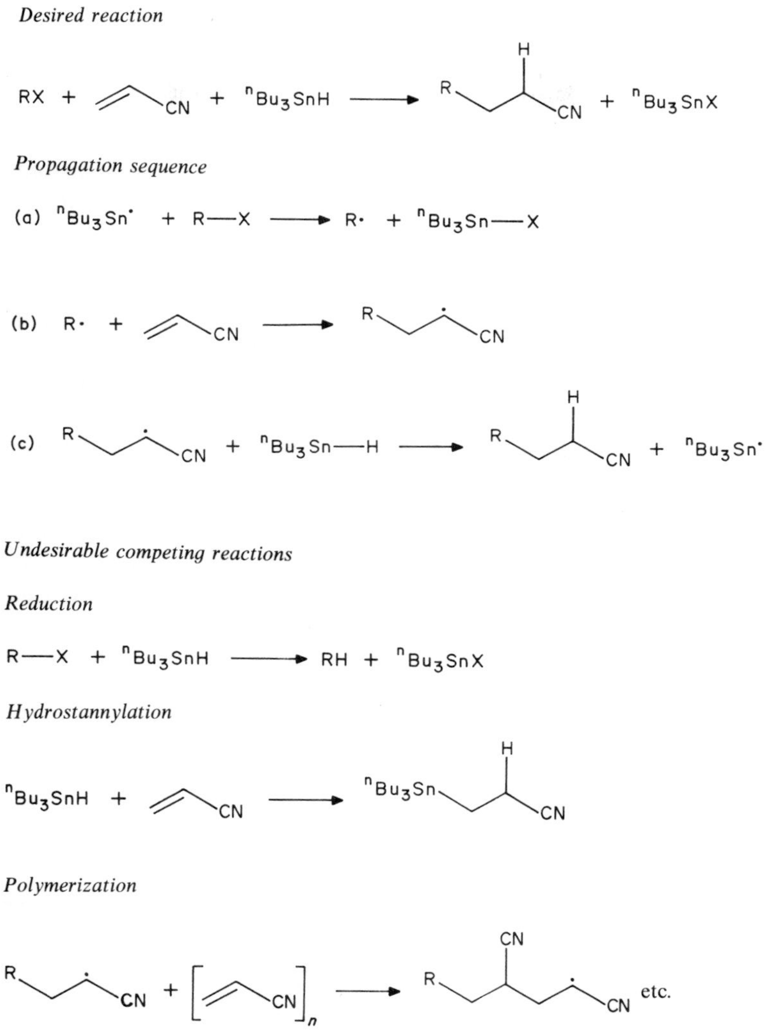

Desired reaction

$$RX + \underset{CN}{\diagup\!\!\diagdown} + {}^nBu_3SnH \longrightarrow R\underset{CN}{\diagdown\!\!\diagup\overset{H}{\diagdown}} + {}^nBu_3SnX$$

Propagation sequence

(a) $\quad {}^nBu_3Sn^{\bullet} + R{-}X \longrightarrow R^{\bullet} + {}^nBu_3Sn{-}X$

(b) $\quad R^{\bullet} + \underset{CN}{\diagup\!\!\diagdown} \longrightarrow R\underset{CN}{\diagdown\!\!\diagup^{\bullet}}$

(c) $\quad R\underset{CN}{\diagdown\!\!\diagup^{\bullet}} + {}^nBu_3Sn{-}H \longrightarrow R\underset{CN}{\diagdown\!\!\diagup\overset{H}{\diagdown}} + {}^nBu_3Sn^{\bullet}$

Undesirable competing reactions

Reduction

$$R{-}X + {}^nBu_3SnH \longrightarrow RH + {}^nBu_3SnX$$

Hydrostannylation

$${}^nBu_3SnH + \underset{CN}{\diagup\!\!\diagdown} \longrightarrow {}^nBu_3Sn\underset{CN}{\diagdown\!\!\diagup\overset{H}{\diagdown}}$$

Polymerization

$$R\underset{CN}{\diagdown\!\!\diagup^{\bullet}} + \left[\underset{CN}{\diagup\!\!\diagdown}\right]_n \longrightarrow R\underset{CN}{\diagdown\!\!\diagup\overset{CN}{\diagdown}\diagup^{\bullet}} \quad etc.$$

SCHEME 1.8

is a slower process than its addition to the electron deficient olefin. Hydrostannylation of the olefin and recovery of the chloride would result. In the case of alkyl bromides, halogen atom abstraction and addition to the olefin are equally attractive possibilities for the tin radical, and occur at similar rates. The weaker carbon–iodine bond of an alkyl iodide, however, ensures that formation of the iodostannane dominates over addition and that alkyl radicals R˙ are constantly regenerated for chain reaction.

(2) *Relative concentrations and modes of addition.* In order to form the desired carbon–carbon bond the experiment must be conducted in a manner to ensure that the alkyl radical R˙ reacts with a molecule of acrylonitrile and not with a molecule of tin hydride. As both of these reactions occur at similar rates, the desired pathway is controlled statistically by slow addition of tri-n-butylstannane to the reaction vessel containing both the iodide and an excess of acrylonitrile. An obvious danger in this approach is that the initial product radical $RCH_2\dot{C}HCN$ is capable of consecutive additions to acrylonitrile leading to polymer formation and this may occur if the excess of acrylonitrile used is too large. Once again, the kinetic knowledge that hydrogen atom transfer from tri-n-butyltin hydride to a radical, irrespective of the nature of the substituents on that radical, is some 10 000 times faster than the addition of this radical to acrylonitrile, determines that an optimum ratio is to use a 100-fold excess of acrylonitrile relative to tri-n-butylstannane [13].

1.5 THE INITIATION CEREMONY

The vision of the organic chemist peering into an inert reaction vessel, muttering an incantation and then adding a small pinch of solid to trigger off a vigorous reaction borders almost on the alchemical, and has certainly not enhanced the credibility of practioners of the radical art among more ionically minded colleagues.

In the ideal world, only a single free radical is required to initiate a transcendentally long propagation sequence. In practice, however, particularly when a more energetically demanding collision is required between the radical and the neutral substrate, as in a bimolecular homolytic substitution, some encounters are unsuccessful and the diffusion controlled combination and radical disproportionation reactions can now enter the scene and terminate chains. This is readily appreciated by comparing the dehalogenation of alkyl iodides and alkyl chlorides by tri-n-butylstannane.

Alkyl iodides rarely require initiation. A trace of free iodine or a molecule of oxygen is sufficient to generate a tin radical by homolytic cleavage of the tin–hydrogen bond. Thereafter, almost every encounter with an alkyl iodide is productive, the propagation sequence turns over smoothly, and an inordinately long chain length ensues. By way of contrast, a much greater number of the encounters between an alkyl chloride and a tin radical fail to induce the required homolysis of the carbon–chlorine bond. Complete reaction is only assured by controlled generation of further tin radicals using an initiation sequence such as the addition of small quantities of azobisisobutyronitrile (AIBN) (Scheme 1.9). The addition of a thermal initiator, all in one portion, albeit of only 5–10 mol%, is not a wise idea since the ensuing flood of radicals can combine and disproportionate in a counter-productive way; and slow addition in solution over a period of time is much more effective. Unfortunately, many literature applications of known free radical chain reactions have not sought to achieve the ideal situation of a single radical in the reaction vessel at any given moment in time. Experiments in which substrate, reagent(s) and initiator have been premixed prior to heating should always therefore be viewed with some suspicion, particularly if the yield is low.

Initiation sequences

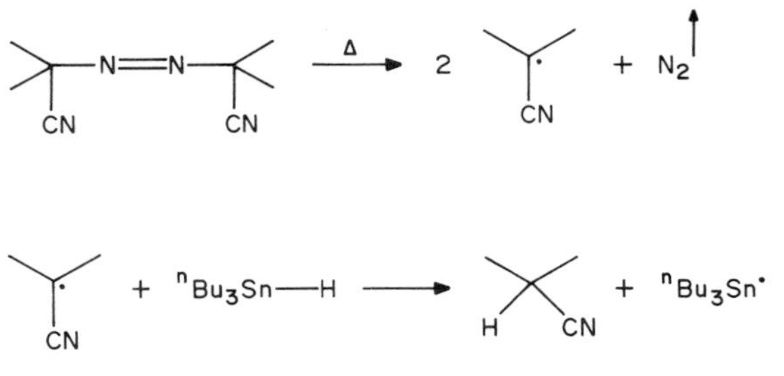

SCHEME 1.9

The correct choice of initiator is generally decided by the operating temperature and hence by the appropriate half-life of the decomposition reaction. A further criterion is the selection of an initiating radical which has the correct character for the reaction in hand. Cyanoisopropyl radicals are relatively docile, although perfectly capable of abstracting a hydrogen atom from the weak triorganostannane Sn—H bond. More powerful electrophilic alkoxyl radicals would be required to induce cleavage of the α-C—H bond of amines and ethers or the alkyl C—H bond of esters, whereas abstraction from the acyl group of an ester would require a nucleophilic alkyl radical.

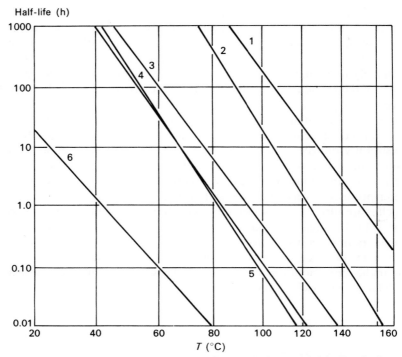

FIG. 1.3. Half-lives for decomposition of typical thermal initiators (values are approximate since rates may vary with solvent). (1) Di-t-butyl peroxide; (2) t-butyl perbenzoate; (3) dibenzoyl peroxide; (4) $S_2O_8^{2-}$; (5) azobisisobutyronitrile (AIBN); (6) di-t-butyl peroxalate. (Reprinted with permission from C. Walling, *Tetrahedron*, **41**, 3890 (1985).)

Some representative examples are shown in Fig. 1.3 and in Table 1.1

For reactions with particularly low temperature requirements, thermal initiation is obviously impractical. In such cases, photoinitiation may prove useful, and AIBN and dibenzoyl peroxide have been shown to be particularly useful [14]. A new development involving initiation by sonication [15] has also been reported.

1.6   STRATEGY AND DESIGN IN SYNTHETIC SEQUENCES INVOLVING CARBON CENTRED RADICAL REACTIONS

1.6.1   Chemoselectivity in functional group interconversions

In the planning and execution of any synthetic sequence involving a polyfunctional molecule, attention must be given to the cosmetic but

TABLE 1.1

Some commonly used radical initiators

Radical initiator	Radical(s) produced	Half-life (h)	at temperature (°C) of
Di-t-butyl peroxalate	$(CH_3)_3C—O^{\cdot}$	12	25
Azobisisobutyronitrile	$(CH_3)_2\overset{\cdot}{C}$ $\quad$ CN	2	80
		0.1	100
t-Butyl perbenzoate	$(CH_3)_3C—O^{\cdot}$ $PhCO_2^{\cdot}$ $\quad Ph^{\cdot}$	20	100
		1	125
Dibenzoyl peroxide	$PhCO_2^{\cdot}$ $\quad Ph^{\cdot}$	2	90
		0.5	100
Di-t-butyl peroxide	$(CH_3)_3C—O^{\cdot}$	200	100
		6.4	130

CAUTION: Diacetyl peroxide and di-t-butyl peroxalate can explode without warning and should therefore be handled with care. Dialkyl peroxydicarbonates (RO—CO—O—O—CO—OR) which decompose faster than dibenzoyl peroxide and undergo rapid self-induced decomposition should also be treated with respect.

fundamental problem of functional group compatibility, and each step scrutinized to determine if protecting groups are required to ensure chemoselective reaction. In this respect, a simple nucleophilic carbon centred radical is a most docile intermediate, preferring, as we have seen, to add to an electron deficient olefin or to remove the weakly bound hydrogen atom of an organostannane. Intermolecular addition to the carbonyl group to give a high energy alkoxyl radical is not a favoured process and protection of this moiety is therefore unnecessary.

An important corollary to this observation is that, unlike nucleophilic carbanions, competing 1,2- and 1,4-addition to an α,β-unsaturated carbonyl group does not occur and only regiospecific conjugate addition of the soft radical is observed (Scheme 1.10).

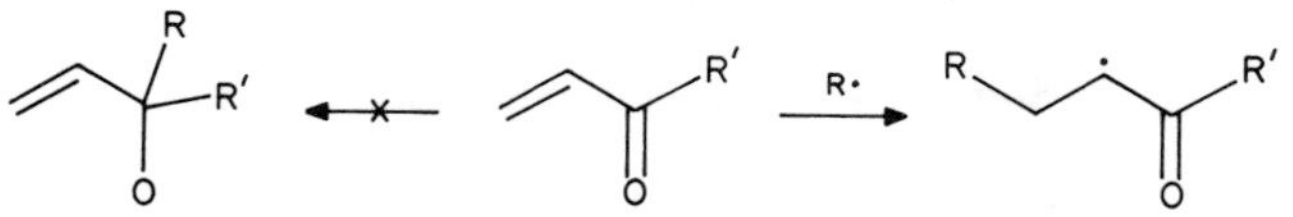

SCHEME 1.10

Protection of the following groups is unnecessary:

$\succ=0$ —OH   —NH$_2$

SCHEME 1.11

Equally improbable reactions include hydrogen atom abstraction from hydroxyl and amino groups which are highly endothermic (Scheme 1.11). Attention should, however, be given to the nature of the vicinal functional group Y adjacent to a carbon centred radical, in order to prevent an undesired elimination reaction (Scheme 1.12).

Elimination reactions triggered by carbon centred radicals

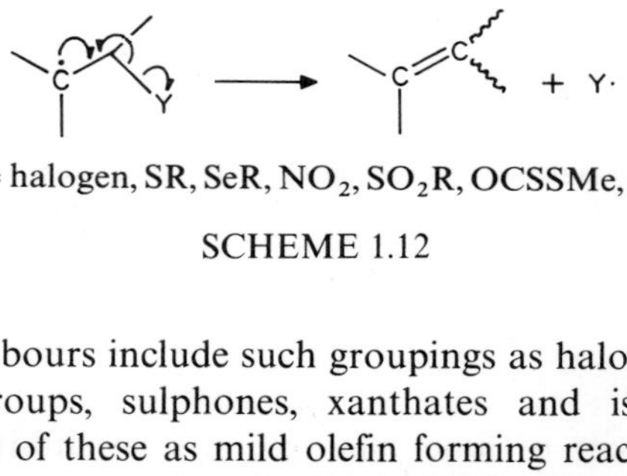

Y = halogen, SR, SeR, NO$_2$, SO$_2$R, OCSSMe, NC

SCHEME 1.12

Dangerous neighbours include such groupings as halogens, sulphides and selenides, nitro groups, sulphones, xanthates and isonitriles, and the preparative aspects of these as mild olefin forming reactions are discussed later (Chapter 3). In contrast to carbanionic eliminations, however, "leaving" groups such as acetate and mesylate are perfectly compatible with a neighbouring carbon centred radical. Intermolecular hydrogen atom abstraction or inter/intramolecular addition to an olefinic double bond can therefore be carried out without fear.

1.6.2   Stereoselectivity in carbon–carbon bond forming reactions

The necessity for highly stereoselective reactions in modern organic synthesis immediately invites the criticism that a carbon centred free radical is a planar

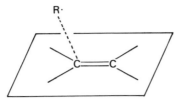

FIG. 1.4. The transition state formed in a radical addition reaction.

entity and hence high stereoselection is impossible in carbon–carbon bond forming reactions. The trite response to this truism is, of course, that the planarity of enolate anions, iminium ions and even free carbocations has not hindered the effective operation of stereoelectronic control elements leading to stereospecific reactions. As Woodward succinctly demonstrated time and again, the presence of a single stereodefined centre is sufficient to induce the creation of proximal neighbours. This effect is particularly manifest in cyclic systems or in reactions proceeding via cyclic transition states where conformational preference is dominant.

In terms of synthetic planning, the feasibility of formation of a crucial carbon–carbon bond by a radical addition reaction is particularly easy to examine. Unlike cationic reactions where attack occurs at the electron rich centre of the double bond, radical additions are generally agreed to proceed via an early and unsymmetrical transition state [16] (Fig. 1.4) in which one of the olefinic carbon atoms is singled out for new bond formation. The favourability of a particular process then follows to a first approximation merely by inspection of orbital overlap of the $R^{\cdot}$ sp^2 centre with a p orbital of the olefin.

The theoretical and practical consequences of this addition mode in intramolecular reactions have been elucidated in a series of elegant investigations by the Beckwith group [17] and are discussed in detail later (Chapter 7). In summary, reactions proceeding under kinetic control occur exclusively in the *exo* mode as defined by Baldwin's rules [18] and are sufficiently faster than competing intermolecular reactions to be useful in synthesis (Schemes 1.13 and 1.14).

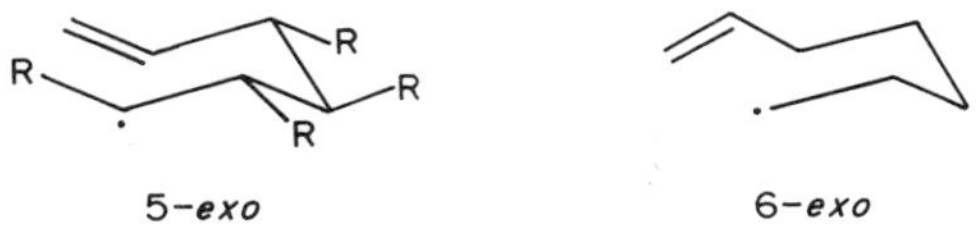

5-*exo*   6-*exo*

SCHEME 1.13

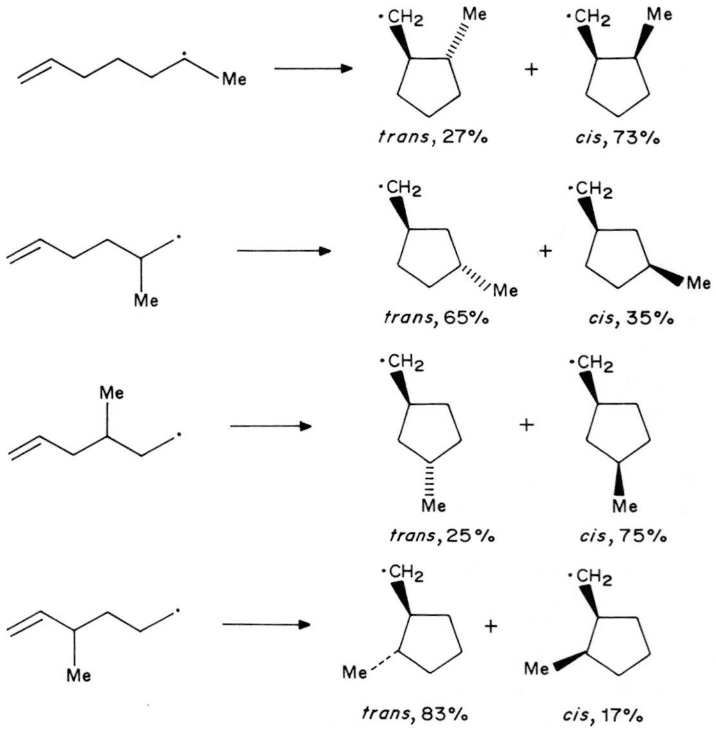

SCHEME 1.14

The preference for a pseudoequatorial array of substituents around the periphery leads to demonstrable stereoselective preference.

The presence of a vicinal centre of defined stereochemistry in a cyclic array is also sufficient to induce stereoselectivity with a preference for the expected *trans* disposition of substituents (Scheme 1.15).

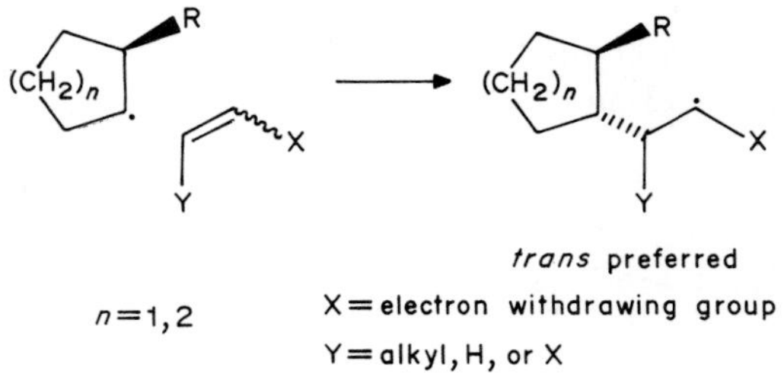

trans preferred

$n = 1, 2$ X = electron withdrawing group
                 Y = alkyl, H, or X

SCHEME 1.15

Several factors influence the degree of stereoselection obtained including the ring size (5 > 6), the degree of unsaturation in the olefinic acceptor (Y = X is best), and the size and electronic nature of the neighbouring substituent R (NHCOCH$_3$ > CH$_3$ > OCH$_3$) [19].

Nevertheless, at the acyclic level, it must be admitted that the stereoselective addition of a reagent X–Y to an olefinic group is most effective in the ionic sense, as a direct consequence of the ability of cationic species to form bridged intermediates (Scheme 1.16):

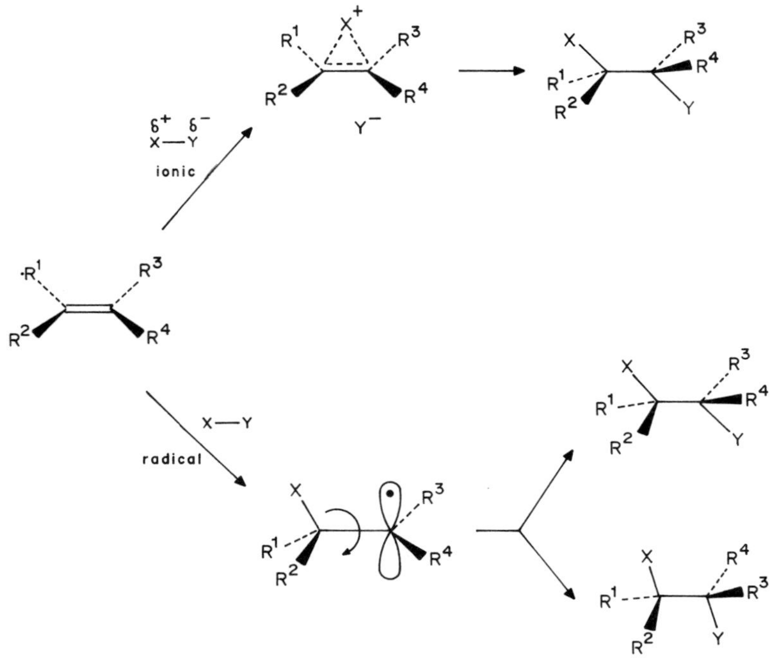

SCHEME 1.16

1.6.3 Retrosynthetic carbon–carbon bond disconnections

In any target oriented synthesis the most important operation in the retrosynthetic sense lies not in the cosmetic interconversion of functional groups but in the selection of a particular carbon–carbon bond or bonds to be broken which lead to a simplified and readily assembled structure.

In traditional organic synthesis the practising chemist is taught to recognize almost instinctively the enormous potential of reactions such as the Diels–Alder cycloaddition, the Aldol condensation, or the Julia or Wittig olefin synthesis, both in the inter- and in the intramolecular modes.

At first sight, the ground rules covering a radical designed carbon–carbon bond forming reaction appear to be somewhat foreign. They are, however, entirely related to the factors governing the controlled generation of a multistep propagative free radical chain sequence (Section 1.4) and are easily summarized in Scheme 1.17, which shows that the most popular and predictably successful strategy lies in the addition of a nucleophilic radical to an electron deficient alkyne or olefin, particularly in the *exo* favoured, kinetically controlled intramolecular mode.

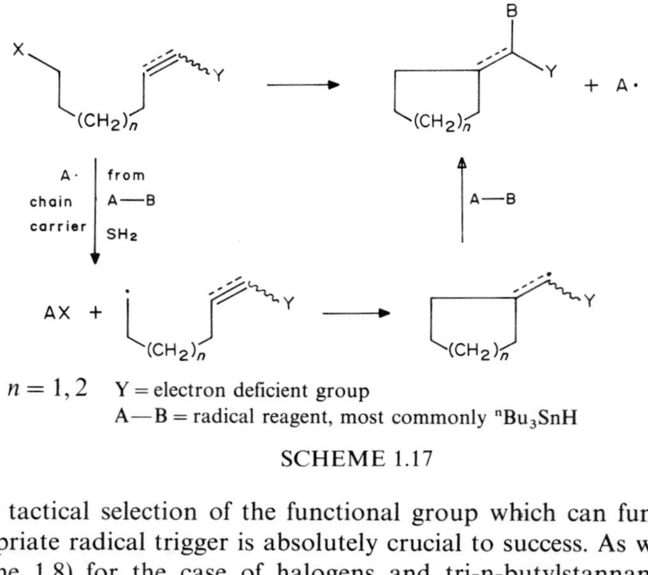

$n = 1, 2$ Y = electron deficient group
A—B = radical reagent, most commonly $^nBu_3SnH$

SCHEME 1.17

The tactical selection of the functional group which can function as an appropriate radical trigger is absolutely crucial to success. As we have seen (Scheme 1.8) for the case of halogens and tri-n-butylstannane, chlorides cannot be used but iodides are perfectly satisfactory.

Accordingly, one of the most often used and predictably successful retrosynthetic steps, in recent years, has been the α-vinyl cleavage reaction. As illustrated for the case of the bicyclic *cis*-hydrindane derivative, two distinct radical cyclization precursors may, in principle, be generated, both of which feature the favoured *exo* cyclization mode (Scheme 1.18).

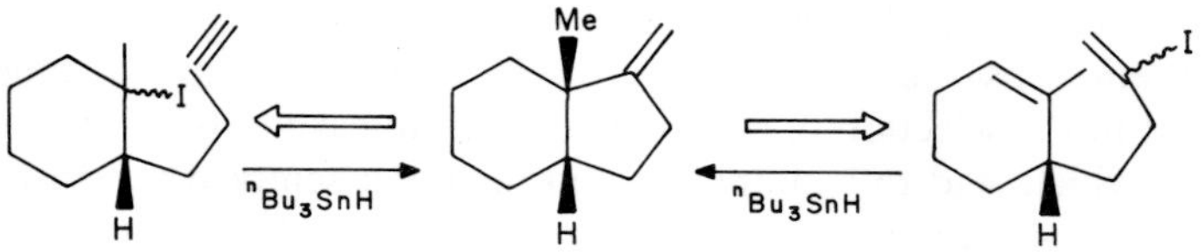

SCHEME 1.18

Target	Precursor	Reagent	Comment	Ref.
		$^nBu_3SnH$	No OH protection necessary	[20]
		$^nBu_3SnH$	Stereochemistry of vinyl iodide is unimportant	[21]
		$^nBu_3SnH$	*Trans* stereochemistry in precursor controls stereospecificity. Two bond mode in a tandem cyclization	[22]
		$^nBu_3SnH$	The "addition–elimination" trick is sometimes useful	[23]
		$^nBu_3SnH$	The radical anomeric effect operates	[24]
		$^nBu_3SnH$	Quaternary centres are easily formed	[25]

SCHEME 1.19

However, the ability to recognize a latent radical disconnection is best achieved by careful study of the highest yielding literature examples. A representative summary of some basic types (which also illustrate some additional points) are shown in Scheme 1.19.

1.7 SOME PRACTICAL GUIDELINES

Some of the following "rules", which should be incorporated into the planning stage of any free radical chain reaction, may be succinctly distilled from the foregoing theoretical treatment.

1.7.1 Initiators

These, if required, should always be added slowly in solution, or portionwise, during the course of the reaction, with due regard for the half-life of their decomposition at the operating temperature of the reaction, and for their chemical ability to carry out a specific transformation, e.g. hydrogen atom abstraction or addition to an alkene. The direct addition of a single portion of AIBN is all too often used in what can only be described as a hopefully optimistic operation.

1.7.2 Solvents

Although polar effects in radical reactions are much less important than in ionic cases and consequently allow a much wider choice in general terms, some specific restrictions should be noted. Self evidently, halogenated solvents should be avoided for reactions involving tin, silicon and phosphorus centred radicals as intermediates. Less often appreciated, particularly when reactions are run at relatively high dilution, is the fact that alkyl radicals will add fairly readily to aromatic solvents such as benzene.

1.7.3 The correct order of addition of reagents

As we have seen, it cannot be sufficiently stressed that, for reactions proceeding under the guidelines of kinetic control, the mode of addition may prove to be absolutely crucial; not only in terms of optimizing the yield of desired product, but even in terms of forming it at all. A particularly striking example [26] is found in the reaction of the α-epoxythiocarbonyl imidazolide derivative (Scheme 1.20).

Irrespective of the mode of addition, the capture of the tin radical by the sulphur atom is followed by a series of fragmentation reactions which result in stereoelectronically controlled C—O bond cleavage of the epoxide, to form an allylic alkoxyl radical as the key intermediate. By addition of the substrate to tri-n-butyltin hydride, hydrogen atom transfer from the latter is effectively controlled and the rearranged allylic alcohol is formed. If, however, tri-n-butylstannane is added to the substrate then the allylic alkoxyl radical has time to undergo a further rearrangement involving β-scission prior to hydrogen atom capture from the stannane, and formation of the bridged bicyclic ketone becomes dominant.

At the present time, it is not uncommon to find in an experimental section, that, for example, premixing of substrate and tri-n-butylstannane has been followed by reflux until the starting material has been consumed. In many cases such a procedure may not be the most efficient method for obtaining a high yield of the desired product.

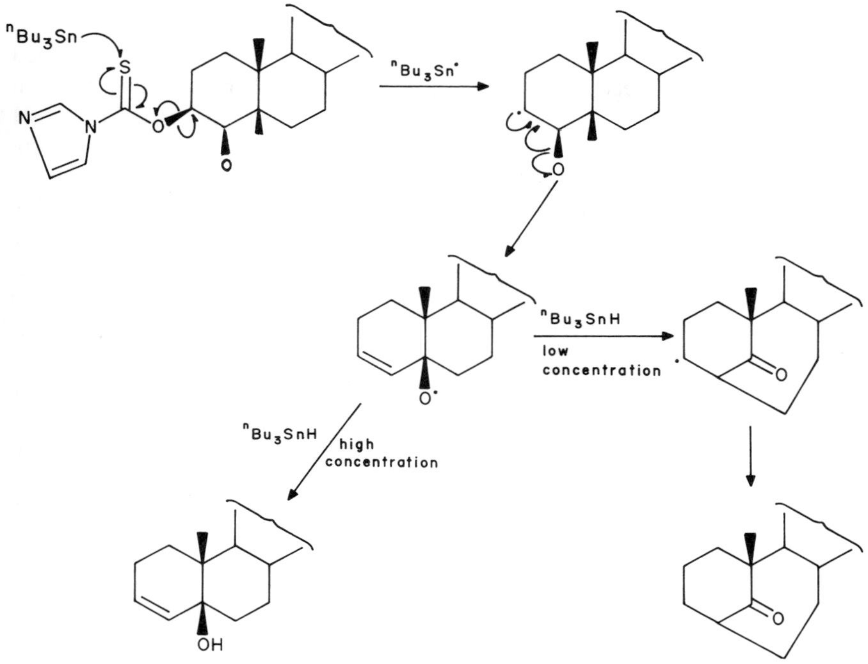

SCHEME 1.20

In those reactions where a multiple series of events is to be followed by hydrogen atom capture, it follows that the concentration of the donor should be relatively low. A useful strategy in such cases has been developed by Corey [27] wherein tri-n-butyltin chloride can be employed in catalytic amounts and regenerated by *in situ* sodium borohydride reduction. This technique, provided that other functional groups in the molecule are not reduced by borohydride, also simplifies the problem of removal of organostannyl by-products.

1.7.4 Removal of organotin residues

Since many of the reactions described in this book involve the use of tri-n-butyltin hydride, it is appropriate, at the outset, to address the important practical problem of separating a desired product from the organostannane residue which has been produced. Such work up procedures can often be applied, although it should be mentioned that reactions producing tri-n-butyltin phenylselenide pose no problems and are simple to purify.

Method A

A simple solution [26] consists of adding carbon tetrachloride to the cooled reaction mixture and monitoring the disappearance of any residual tri-n-butyltin hydride by the decrease in the strong characteristic Sn—H stretch in the infrared spectrum at 1800 cm^{-1}. In this way tri-n-butyltin chloride is produced from the hydride. A dilute solution of iodine in ether is then added until a faint coloration persists indicating that cleavage of any hexabutyldistannane to tri-n-butyltin iodide has occurred. The organic solvent is then removed in *vacuo* and the residue taken up in diethyl ether or ethyl acetate and washed several times with a saturated aqueous solution of potassium fluoride until no more precipitation of the flocculent polymeric tri-n-butyltin fluoride is observed. The use of diethyl ether is recommended since it ensures production of an insoluble precipitate rather than an oil. The separated organic phase may then be dried and solvent removed in the normal manner. A conceptually similar treatment has also been detailed for acyl thiohydroxamate decarboxylations [28].

Method B

A second solution, which has also proved to be a practical proposition, makes use of the essentially lipophilic nature of tin residues [29]. Consequently, simple partitioning between polar solvents such as acetonitrile or wet methanol and pentane can often prove effective for more polar substrates such as carbohydrates or aminoglycoside antibiotics.

Method C

A simple protocol [30] for removal of distannanes and tin halides consists of treatment with 1,5-diazabicycloundecene (DBU). The reaction mixture is diluted with reagent grade (undried) ether (10–20 ml). DBU (0.2 equiv., for 0.1 equiv. hexaalkylditin) is added to the reaction mixture and then titrated with 0.1 M iodine solution. During this time, DBU-hydroiodide precipitates as a white solid. After the iodine colour just persists, the solution is transferred to a short column (SiO$_2$); after elution with ether (30 ml), the solvent is removed. The residue is almost tin free. If dehalogenated products are desired, the procedure is applied following tin hydride (1.2 equiv.) treatment. Excess DBU (1.5 equiv.) is used in order to remove all tin halide. The sequence of addition of DBU followed by titration with iodine solution can be inverted.

The simplest solution of all is, of course, to avoid the use of organo-stannane derivatives, and much recent work involving, for example, acyl thiohydroxamates and organocobalt complexes (*vide infra*) takes this aspect into account. Two particularly pertinent approaches involving the use of industrially acceptable silanes are, however, worthy of special attention.

Thus the introduction of *tris*-(trimethylsilyl)silane [31] and its promotion by Giese [32] as a slower, but effective, hydrogen atom donor than tri-n-butylstannane, offers a useful alternative for larger scale work.

A delightfully simple and ingenious solution to classical free radical chain dehalogenation has been introduced by Roberts [33] and involves the use of standard trialkylsilanes in the presence of 5–10 mol% of thiophenol.

The success of this method relies on the insertion of an additional step into the propagation sequence and, as always, is a reflection of kinetic considerations (Scheme 1.21). Substrates possessing an isolated double bond, to which thiophenyl radicals may add, cannot of course be used.

Propagation sequence

$$RBr + R^1_3Si\cdot \longrightarrow R\cdot + R^1_3SiBr$$
$$R\cdot + PhSH \longrightarrow RH + PhS\cdot$$
$$PhS\cdot + R^1_3SiH \longrightarrow PhSH + R^1_3Si\cdot$$

SCHEME 1.21

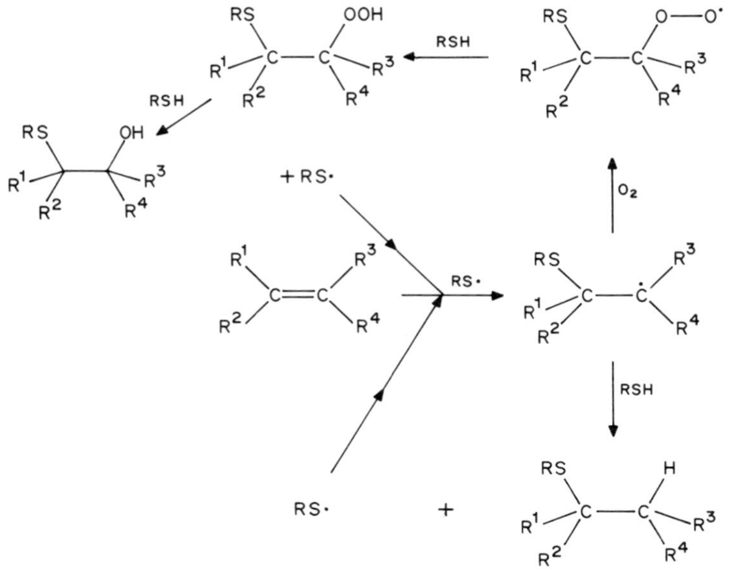

SCHEME 1.22

1.7.5   The role of oxygen in free radical reactions

One of the simplest classical free radical chain reactions is the addition of a mercaptan to an olefin [34]. In the presence of oxygen, however, hydrogen atom abstraction from the thiol by the intermediate carbon centred radical cannot compete and smooth hydroxysulphenylation ensues [35] (Scheme 1.22). In fact, competitive experiments show that reaction of the growing carbon centred radical in polymerization of styrene or methyl methacrylate with the triplet oxygen diradical is some 10 000 times faster than hydrogen donation from the mercaptan. Accordingly, *reactive oxygen should always be excluded from free radical chain reactions unless its incorporation into the product is a desirable feature.*

1.8   OVERVIEW AND PERSPECTIVES

The essence of organic synthesis has always been to achieve chemospecific reactions of predictable regio- and stereochemical outcomes under mild neutral conditions. In an ideal world such "general" reactions could then be universally applied irrespective of the polarity, complexity and nature of the surrounding molecular architecture. The foregoing generalizations have hopefully indicated that the carbon centred free radical, more than any other reactive intermediate, is ideally equipped to perform in this role.

Some recent radical triggers used for carbon–carbon bond formation

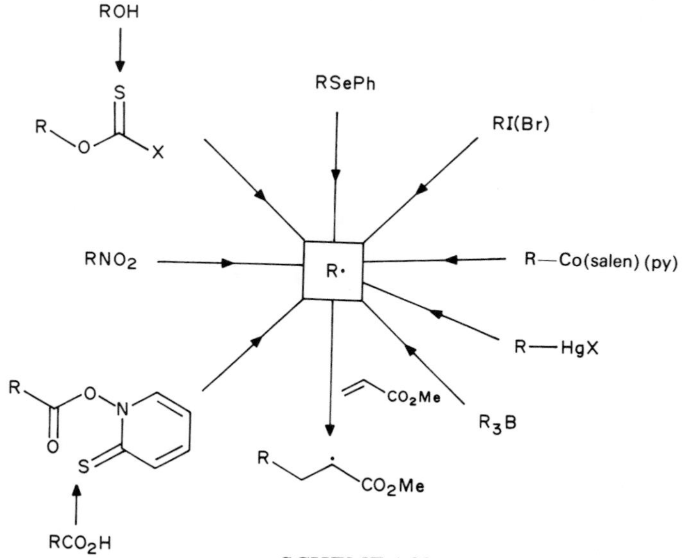

SCHEME 1.23

A wide selection of functional groups, or suitable derivatives thereof, can now be used to generate a carbon centred radical under mild conditions and some of those which have been consistently used both for functional group transformations and in inter- and/or intramolecular carbon–carbon bond forming reactions are illustrated in Scheme 23. Their evolution and development form the subject matter of the following chapters.

The demonstrable virtues of radical chemistry will no doubt continue to attract increasing numbers of synthetic organic chemists in the future and, it is hoped, lead to increased efforts devoted to even milder and environmentally clean methods for the generation and exploitation of these valuable intermediates.

REFERENCES

1. D. H. R. Barton, *ipse dixit*, freely translated from the original "libre mais bien domestiqué". Gif-sur-Yvette, 1981.
2. For discussions see: C. Walling, *Tetrahedron* **41**, 3887 (1985); S. F. Martin, in *Free Radicals* (ed. J. K. Kochi), Vol. 1, p. 493. Wiley-Interscience, New York, 1973.
3. J. K. Kochi (ed.), *Free Radicals*, Vols 1, 2. Wiley-Interscience, New York, 1973. Many of the chapters are organized as a function of the atom containing the unpaired electron.
4. E. J. Corey and E. Hamanaka, *J. Am. Chem. Soc.* **89**, 2758 (1967).
5. B. Giese and G. Kretzschmar, *Chem. Ber.* **116**, 3267 (1983).
6. I. Fleming, *Frontier Orbitals and Organic Chemical Reactions.* Wiley, London, 1976; B. Giese, *Angew. Chem. Int. Edn Engl.* **22**, 771 (1983).
7. B. Giese and H. Horler, *Tetrahedron Lett.* **24**, 3221 (1983); B. Giese and H. Horler, *Tetrahedron*, **41**, 4025 (1985).
8. B. Giese, H. Horler and M. Leising, *Chem. Ber.* **119**, 444 (1986).
9. J. G. Noltes and G. J. M. Van der Kerk, *Chem. Ind.* 294 (1959).
10. H. G. Kuivila, *Synthesis* 499 (1970).
11. D. Griller and K. U. Ingold, *Acc. Chem. Res.* **13**, 317 (1980).
12. B. Giese, *Angew. Chem. Int. Edn Engl.* **24**, 553 (1985).
13. For a detailed quantitative kinetic analysis see: B. Giese, in *Radicals in Organic Synthesis, Formation of Carbon–Carbon Bonds*, pp. 6–12. Pergamon Press, Oxford, 1986.
14. P. S. Engel, *Chem. Rev.* **80**, 99 (1980); C. Walling and M. J. Gibian, *J. Am. Chem. Soc.* **87**, 3413 (1965); R. A. Sheldon and J. K. Kochi, *J. Am. Chem. Soc.* **92**, 4395 (1970).
15. E. Nakamura, D. Machii and T. Inubushi, *J. Am. Chem. Soc.* **111**, 6849 (1989).
16. M. J. S. Dewar and S. Olivella, *J. Am. Chem. Soc.* **100**, 5290 (1978); H. Fujimoto, S. Yamabe, T. Minato and K. Fukui, *J. Am. Chem. Soc.* **94**, 9205 (1972).
17. A. L. J. Beckwith and C. H. Schiesser, *Tetrahedron* **41**, 3925 (1985); A. L. J. Beckwith, G. Phillipou and A. K. Serelis, *Tetrahedron Lett.* **22**, 2811 (1981); A. L. J. Beckwith and G. F. Meijs, *J. Chem. Soc. Perkin Trans.* **2**, 1535 (1979).
18. J. E. Baldwin. *J. Chem. Soc. Chem. Commun.* 734 (1976).
19. R. Henning and H. Urbach, *Tetrahedron Lett.* **24**, 5343 (1983); B. Giese, K. Heuck, H. Lenhardt and U. Lüning, *Chem. Ber.* **117**, 2132 (1984); B. Giese, H. Harnisch and U. Lüning, *Chem. Ber.* **118**, 1345 (1985).

20. G. Stork and N. H. Baine, *J. Am. Chem. Soc.* **104**, 2321 (1982).
21. N. N. Marinovic and H. Ramanathan, *Tetrahedron Lett.* **24**, 1871 (1983).
22. D. P. Curran and D. M. Rakiewicz, *Tetrahedron*, **41**, 3943 (1985).
23. Y. Ueno, K. Chino and M. Okawara, *Tetrahedron Lett.* **23**, 2575 (1982).
24. J. Dupuis, B. Giese, J. Hartung, M. Leising, H.-G. Korth and R. Sustmann, *J. Am. Chem. Soc.* **107**, 4332 (1985).
25. N. Ono, H. Miyake, A. Kamimura, I. Hamamoto, R. Tamura and A. Kaji, *Tetrahedron* **41**, 4013 (1985).
26. D. H. R. Barton, R. S. Hay-Motherwell and W. B. Motherwell, *J. Chem. Soc. Perkin Trans.* **1**, 2363 (1981), and references therein.
27. E. J. Corey and J. W. Suggs, *J. Org. Chem.* **40**, 2554 (1975).
28. D. H. R. Barton, W. B. Motherwell and D. Crich, *Tetrahedron* **41**, 3901 (1985).
29. J. M. Berge and S. M. Roberts, *Synthesis*, 471 (1979).
30. D. P. Curran and C.-T. Chang, *J. Org. Chem.* **54**, 3140 (1989).
31. C. Chatgilialoglu, D. Griller and M. Lesage, *J. Org. Chem.* **53**, 3641 (1988).
32. B. Giese, B. Kopping and C. Chatgilialoglu, *Tetrahedron Lett.* **30**, 681 (1989).
33. R. P. Allen, B. P. Roberts and C. R. Willis, *J. Chem. Soc. Chem. Commun.* 1387 (1989).
34. F. W. Stacey and J. F. Harris Jr, *Org. Reactions* **13**, 150 (1963); R. M. Kellog, *Methods Free Radical Chem.* **2**, 1 (1969).
35. M. S. Kharasch, W. Nudenberg and G. J. Mantell, *J. Org. Chem.* **16**, 524 (1951).

—2—

Books and Review Articles

The enormous growth of interest in the practical applications of preparative free radical reactions has been appropriately reflected in recent years by a variety of useful books and review articles, and by an ultimate accolade in the form of two volumes in the Houben-Weyl series.

It is therefore appropriate, at this stage, to include some of the more valuable sources for consultation.

2.1 SOME EARLIER BASIC TEXTS AND REVIEWS (pre-1980)

Many of the older texts, when revisited, with the added knowledge accumulated in intervening years, are valuable sources of inspiration. The two volumes of *Free Radicals* by Kochi (ref. [9]) should, in particular, be considered as a *vade mecum* of any radical chemist's library

1. C. Walling, *Free Radicals in Solution*. John Wiley, New York, 1957.
2. G. Sosnovsky, *Free Radical Reactions in Preparative Organic Chemistry*. MacMillan, London, 1964.
3. W. A. Pryor, *Free Radicals*. McGraw Hill, New York, 1966.
4. G. B. Gill, *Synthetic Uses of Free Radicals in 'Modern Reactions in Organic Synthesis'* (ed. C. J. Timmons), Chap. 4, p. 90. Van Nostrand Rheinhold, London, 1970.
5. E. S. Huyser, *Free Radical Chain Reactions*. Wiley-Interscience, New York, 1970.
6. K. U. Ingold and B. P. Roberts, *Free-Radical Substitution Reactions, Bimolecular Homolytic Substitutions (S_H2 Reactions) at Saturated Multivalent Atoms*. Wiley-Interscience, New York, 1971.
7. C. S. Walling and E. S. Huyser, *Org. Reactions* **13**, 91 (1963).
8. M. Julia, Free radical cyclisations, *Acc. Chem. Res.* **4**, 386 (1971).
9. J. K. Kochi (ed.), *Free Radicals*, Vols 1, 2. Wily-Interscience, New York, 1973.
10. D. C. Nonhebel and J. C. Walton, *Free Radical Chemistry*. Cambridge University Press, Cambridge, 1974.
11. D. I. Davies and M. J. Parrot, *Free Radicals in Organic Synthesis*. Springer-Verlag, Berlin, 1978.
12. A. L. J. Beckwith and K. U. Ingold, *Rearrangements in Ground and Excited States* (ed. P. de Mayo), Vol. 1, p. 161. Academic Press, New York, 1980.
13. D. Griller and K. U. Ingold, Free radical clocks, *Acc. Chem. Res.* **13**, 317 (1980).

2.2 GENERAL BOOKS AND REVIEWS (Post-1980)

14. B. Giese (ed.), Selectivity and synthetic applications of radical reactions, *Tetrahedron Symposia* 22, *Tetrahedron* **41** (1985).
15. D. J. Hart, Free-radical carbon–carbon bond formation in organic synthesis, *Science* **223**, 883 (1984).
16. B. Giese, *Radicals in Organic Synthesis, Formation of Carbon–Carbon Bonds*. Pergamon Press, Oxford, 1986.
17. M. Ramiah, Radical reactions in organic synthesis, *Tetrahedron Report* 223, *Tetrahedron* **43**, 3541 (1987).
18. J. M. Tedder, Which factors determine the reactivity and regioselectivity of free radical substitution and addition reactions? *Angew. Chem. Int. Edn* **21**, 401 (1982).
19. D. P. Curran, The design and application of free radical chain reactions in organic synthesis, *Synthesis* **417**, 489 (1988).
20. M. Regitz and B. Giese, Carbon radicals, in *Houbern-Weyl, Methoden der Organischen Chemie*, Vols 1, 2, E19a. Georg Thieme Verlag, Stuttgart, 1989.

2.3 RECENT BOOKS AND ARTICLES DEALING WITH MORE SPECIFIC ASPECTS OF FREE RADICAL REACTIONS

21. M. Pereyre, J. P. Quintard and A. Rahm, *Tin in Organic Synthesis*. Butterworths, London, 1986.
22. W. P. Neumann, Tri-n-butyltin hydride as reagent in organic synthesis, *Synthesis* 665 (1987).
23. A. L. J. Beckwith, Regioselectivity and stereoselectivity in radical reactions, *Tetrahedron* **37**, 3073 (1981).
24. H. G. Viehe, Z. Janousek, R. Merenyi and L. Stella, The captodative effect, *Acc. Chem. Res.* **18**, 148 (1985).
25. J. Barluenga and M. Yus, Free radical reactions of organomercurials, *Chem. Rev.* **88**, 487 (1988).
26. B. Giese, Synthesis with radicals, C—C bond formation *via* organotin and organomercury compounds, *Angew. Chem. Int. Edn Engl.* **24**, 553 (1985).
27. G. A. Russell, Free radical chain reactions involving alkyl- and alkenylmercurials, *Acc. Chem. Res.* **22**, 1 (1989).
28. R. Scheffold, S. Abrecht, R. Orlinski, H.-R. Ruf, P. Stamouli, O. Tinembart, L. Walder and C. Weymoth, Vitamin B_{12}-mediated electrochemical reactions in the synthesis of natural products, *Pure Appl. Chem.* **59**, 363 (1987).
29. G. Pattenden, Cobalt mediated radical reactions in organic synthesis, *Chem. Soc. Rev.* **17**, 361 (1988).
30. T. G. Back, Radical reactions of selenium compounds, in *Organoselenium Chemistry* (ed. D. Liotta), p. 325. Wiley-Interscience, New York, 1987.
31. W. Hartwig, Modern methods for radical deoxygenation of alcohols, *Tetrahedron* **39**, 2609 (1983).
32. D. H. R. Barton and W. B. Motherwell, New and selective reactions and reagents, in *Organic Synthesis Today and Tomorrow* (ed. B. M. Trost). Pergamon Press, Oxford, 1981.
33. D. Crich, *O*-acyl thiohydroxamates: new and versatile sources of alkyl radicals for use in organic synthesis, *Aldrichimica Acta* **20**, 35 (1987).
34. D. Crich and L. Quintero, Radical chemistry associated with the thiocarbonyl group, *Chem. Rev.* **89**, 1413 (1988).

—3—

Substitution Reactions

3.1 DEHALOGENATION (RX → RH, X = halogen)

The reductive removal of a halogen atom by a tin hydride is now considered to be a standard reaction in the armoury of the synthetic chemist. The pioneering observations of Van der Kerk [1], followed by the detailed work of Kuivila [2], have demonstrated that this simple chain reaction is extraordinarily efficient. The formation of the strong tin–halogen bond and the rapid hydrogen atom abstraction by the alkyl radical from the weak tin–hydrogen bond lead in a formidable thermodynamic combination to constant regeneration of the chain carrying tin centred radical and hence to long chain lengths (Scheme 3.1). Some generally useful ground rules governing [2–4] this process are given in the following subsections.

Propagation sequence

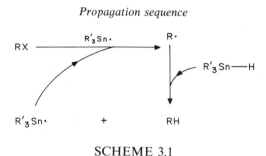

SCHEME 3.1

3.1.1 Selection of the organostannane

Although tri-n-butylstannane is almost universally used, it is in fact one of the less reactive hydrides, reactivity following the order:

$$^nBu_3SnH < {}^nBu_2SnH_2 < Ph_3SnH < Ph_2SnH_2.$$

This factor may become important when possibilities exist for rearrangement of the intermediate radical R·, as in the classic case of cyclopropyl carbinyl radical ring opening [5].

major   Br   major

3.1.2 Selection of the halogen atom

Bromides and iodides tend to react spontaneously with tri-n-butyltin hydride while chlorides may require heating or initiation by azobisisobutyronitrile (AIBN) and/or ultraviolet irradiation. Fluorides are generally considered to be inert. Chemoselective differentiation in a polyhalogenated molecule therefore becomes possible, as shown [6].

97%

3.1.3 Choice of solvent and relative concentrations

Aromatic solvents such as benzene or toluene have been most often used although a wide range of polarities ranging from cyclohexane through diethyl ether and tetrahydrofuran to methanol have also been reported. These provide a convenient choice of operating temperature and, more importantly, in the case of possible competing radical rearrangement (vide supra) allow the selection of addition mode to favour high or low concentrations of the tin hydride.

3.1.4 Reactivity and selectivity as a function of substrate structure

Variation of the alkyl or aryl group follows the expected trend of reactivity:

tertiary > secondary > primary > aryl or vinyl.

Thus formation of higher energy aryl or vinyl radicals accordingly requires a higher operating temperature, and selection of the iodide or bromide is now mandatory.

For a given dehalogenation passing through a planar radical intermediate, little stereoselectivity is to be expected nor indeed is it observed in simple cases. Steric approach control in the hydrogen atom transfer step can, however, lead to highly stereoselective reactions as shown for the case of the penicillin [7] and steroid [8] derivatives where the overall shape of the substrate dictates a unique face of the molecule for the product forming step to occur.

In such cases, the use of tri-n-butyltin deuteride may be used to advantage for highly stereoselective labelling.

Electronic effects on the configurational stability of radicals have, however, been noted in the case of the radical anomeric effect [9,10].

In general terms, the organostannanes are much more tolerant of a wide range of surrounding functionality than their aluminium counterparts. This high degree of chemoselectivity is borne out by the vast numbers of dehalogenation reactions which have been carried out over the years. Some further examples, from various classes of organic families, have been collected in Tables 3.1–3.3 by way of illustration.

TABLE 3.1

Chemoselective reduction of aliphatic and alicyclic halides by organostannanes

Substrate	Halogen	Conditions	Product (Z = H) yield (%)	Ref.
	Z = Br	$^{n}Bu_3SnH$, AIBN, THF, 50°C, 8 h	80	[11]
	Z = Br	$^{n}Bu_3SnH$	—	[12]
	$Z^1 = Z^2 = H$, $Z^3 = Cl$ $Z^1 = Z^3 = H$, $Z^2 = Cl$ $Z^1 = Cl$, $Z^2 = Z^3 = H$	$^{n}Bu_3SnH(D)$	— — —	[13]
	X = F, Z = Br X = OMe, Z = Br	$^{n}Bu_3SnH$ $^{n}Bu_3SnH$	73 81	[14] [15]
	Z = Br	$^{n}Bu_3SnH$, AIBN, benzene	97	[16]
	Z = I	$^{n}Bu_3SnH$	40	[17]

TABLE 3.1 *Continued*

Substrate	Halogen	Conditions	Product (Z = H) yield (%)	Ref.
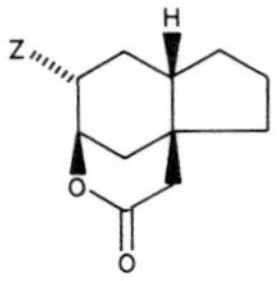	Z = I	$^nBu_3SnH$	> 87	[18]
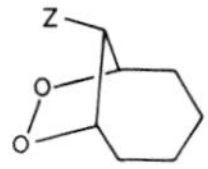	Z = Br	$^nBu_3SnOSn^nBu_3$, $(MeSiHO)_n$	50	[19]
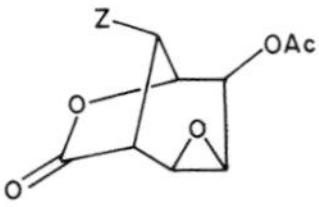	Z = Br	$^nBu_3SnD$	72	[20]
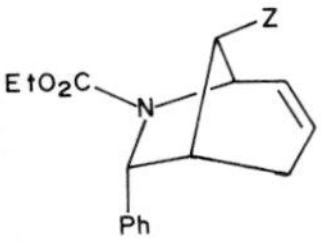	Z = Br	$^nBu_3SnH$	88	[21]
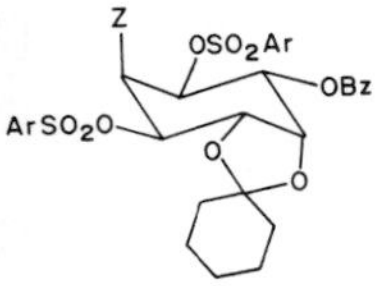	Z = Cl	$^nBu_3SnH$	79	[22]
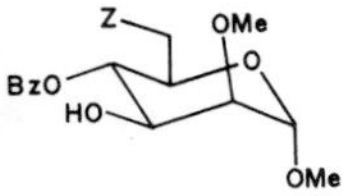	Z = Br	$^nBu_3SnH$	85	[23]
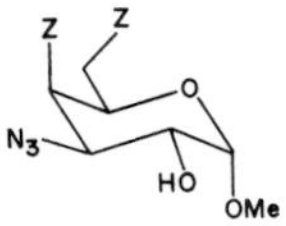	Z = Cl	$^nBu_3SnH$, 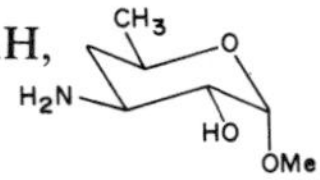	84	[24]
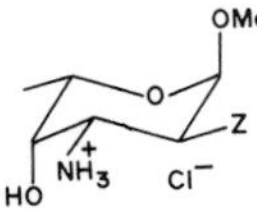	Z = I	$^nBu_3SnH$	65	[25]

TABLE 3.1 *Continued*

TABLE 3.1 *Continued*

Substrate	Halogen	Conditions	Product (Z = H) yield (%)	Ref.
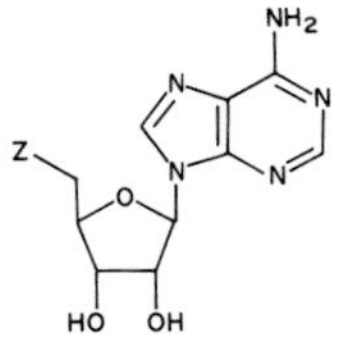	Z = Cl	$^nBu_3SnH$, AIBN	51	[26]
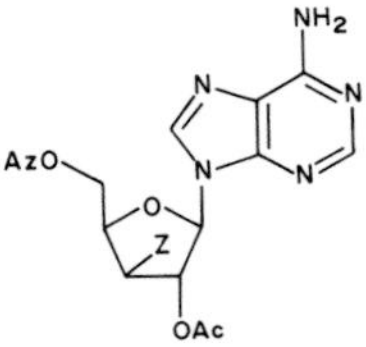	Z = Br	$^nBu_3SnH$	80	[27]

TABLE 3.2

Selective halogen atom removal in polyhalogenated substrates

Substrate	Conditions	Product	Yield (%)	Ref.
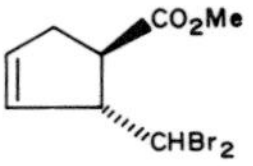	1.0 mol equiv. neat $^nBu_3SnH$, 25°C	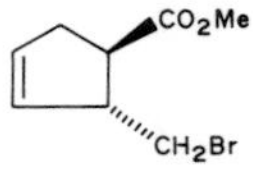	86	[28]
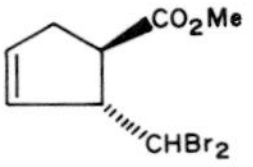	2.0 mol equiv. $^nBu_3SnH$, petroleum ether, heat	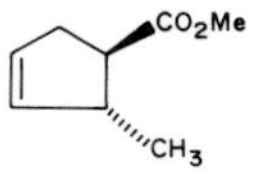	87	[28]

Z—HN—CH—CCl$_3$
         |
    O=Z(OMe)$_2$
 	
 Z = CHO, CO$_2$CH$_2$CH$_3$, CO$_2$CH$_2$Ph
	1.0 mol equiv. $^nBu_3SnH$, benzene, 80°C	ZHN—CH—CHCl$_2$
		         |
		    P(OMe)$_2$
		   O
			75–98	[29]

Z—HN—CH—CCl$_3$
         |
    O=Z(OMe)$_2$
	2.0 mol equiv. $^nBu_3SnH$, benzene, 80°C	ZHN—CH—CH$_2$Cl
		         |
		    P(OMe)$_2$
		   O
			72–97	[29]

TABLE 3.2 *Continued*

Substrate	Conditions	Product	Yield (%)	Ref.
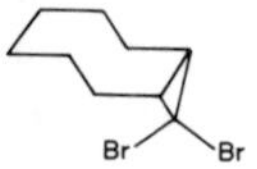	NaBH$_4$, ^{n}Bu$_3$SnCl (catalyst)	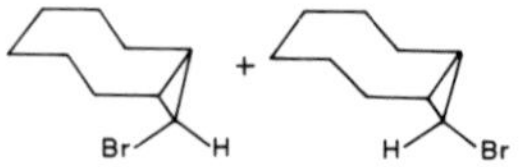 73 : 27		[30]
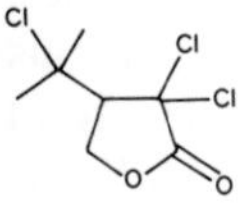	2.0 mol equiv. ^{n}Bu$_3$SnH, AIBN, 80°C, 12 h	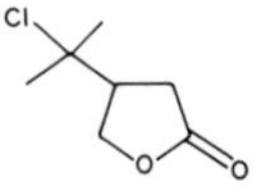	59	[31]
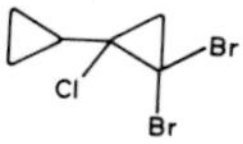	2.0 mol equiv. ^{n}Bu$_3$SnH	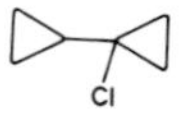	75	[32]

TABLE 3.3

Some typical reduction conditions for vinylic and aromatic halides

Substrate (Z = halogen)	Halogen	Conditions	Product (Z = H) yield (%)	Ref.
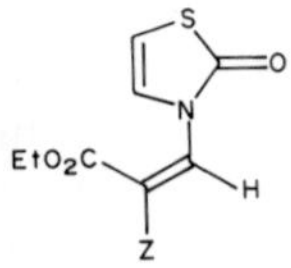	Z = Br	^{n}Bu$_3$SnH, THF 25°C, 72 h	35	[35]
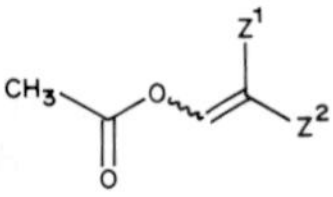	Z^1 = Z^2 = Cl	^{n}Bu$_3$SnH, *hv*	80 (Z^1 = H, Z^2 = Cl)	[34]
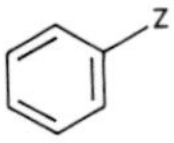	Z = I	Neat ^{n}Bu$_3$SnH, 100–120°C	90	[2]
	Z = Br	Neat ^{n}Ph$_3$SnH, 150°C	90	
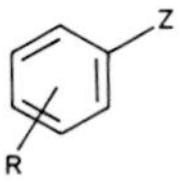	Z = Br R = *m*-, *p*-alkyl; *m*-, *p*-Cl, F, OMe, CN	Et$_3$SnH, *hv*, cyclohexane, 25°C	80–95	[33]

General procedure for dehalogenation reactions (RX → RH) without rearrangement [2]

Dropwise addition of a solution of the substrate (containing 2–5 mol % of initiator if necessary) to a stirred solution of the organostannane maintained at a suitable temperature under an inert atmosphere is followed by aliquot monitoring of the reaction until completion is indicated. Work up involving removal of tin residues may then be carried out by one of the methods indicated in Section 1.7.4.

Photochemical initiation can prove to be particularly useful in several cases, as illustrated below for the room temperature production of aryl radicals [33] and the case of a vinylic chlorine atom replacement where the use of chemical initiators led to substrate and product polymerization [34]. Some further examples are collected in Table 3.3.

Photochemically induced organostannane dehalogenation of 4-bromotoluene—a typical procedure

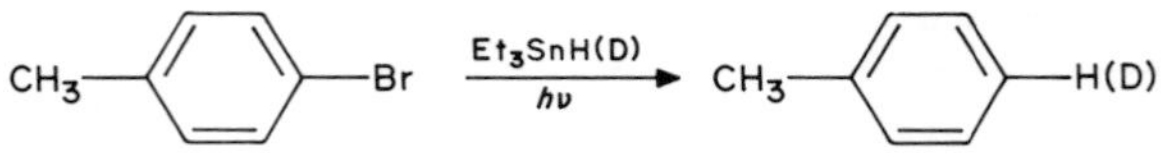

$$CH_3-\!\!\!\bigcirc\!\!\!-Br \xrightarrow[h\upsilon]{Et_3SnH(D)} CH_3-\!\!\!\bigcirc\!\!\!-H(D)$$

Triethyltin hydride (32.6 ml, 200 mmol) was added to 4-bromotoluene (25.66 g, 150 mmol) in a two necked quartz flask and the resulting solvent free mixture irradiated under an argon atmosphere, with magnetic stirring, at room temperature using a 125 W high pressure mercury lamp. After completion (GLC control of aryl bromide) toluene was distilled from the reaction mixture. Traces of tin hydride can be removed by fractionation through a short column. Yield of pure product: 9.9 g (72%), b.p. 110°C/760 mmHg.

Work up of cognate preparations and difficulties in removal of excess tin hydride can be overcome by its decomposition with dichloroacetic acid followed by fractionation.

Replacement of triethyltin hydride by the corresponding deuteride gives 4-deuterotoluene of high isotopic purity.

Translated from ref. [33] with permission.

Preparation of 2-acetoxy-1-chloroethene

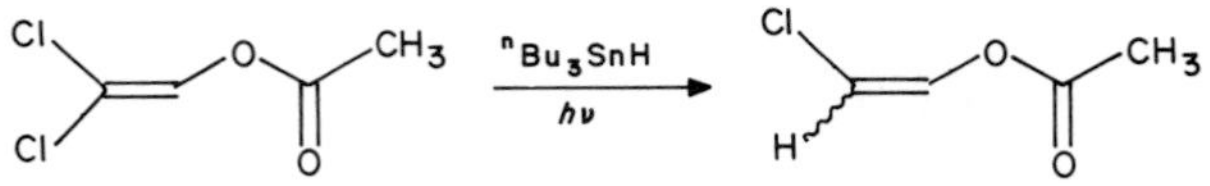

Tri-n-butyltin hydride (33.25 g, 0.114 mol) was added with stirring over 1 h under a nitrogen atmosphere at 0°C to 2-acetoxy-1,1-dichloroethene (17.5 g, 0.113 mol) under irradiation by a 125 W mercury lamp. Stirring was continued for 0.5 h after the addition. Distillation of the reaction mixture yielded 2-acetoxy-1-chloroethane (9 g, 80%), b.p. 71–72°C/98 mmHg, $E:Z$ ratio approximately 1:3.

Translated from ref. [33] with permission.

3.2 DEOXYGENATION REACTIONS (ROH → RH) [36]

3.2.1 Introduction. The development of the thiocarbonyl ester method (Barton–McCombie deoxygenation)

The replacement of a hydroxyl group by a hydrogen atom is traditionally solved by ionic displacement of a suitable sulphonate ester with lithium aluminium hydride. In carbohydrates, however, such reactions are often sluggish and/or unsuccessful because of the opposing dipoles of neighbouring β-carbon–oxygen bonds. The realization by Barton [37,38] that radical reactions would be impervious to such an environment led to an ingenious solution based on the radical chain reduction of a variety of readily prepared thiocarbonyl derivatives by tri-n-butyltin hydride.

$X = SMe, Ph, \ \begin{array}{c} \\ N \end{array} , OPh, OC_6F_5, SPh$

Although xanthates (X = SMe), thiobenzoates (X = Ph) and thiocarbonyl imidazolides (X = 1-imidazolyl) were the first derivatives to be investigated, more recent additions include thionocarbonates (X = OPh [39], OMe [40]) and delocalized thionocarbamates (X = 1-pyrollyl [41], X = 1-(1H)pyridin-2-onyl [42]).

In general mechanistic terms, the reaction is considered to proceed via reversible addition of the organostannyl radical to the thiocarbonyl group followed by fragmentation of the intermediate carbon centred radical to give a carbonyl group with concomitant liberation of the derived alkyl radical R·. For the particular case of xanthates, alternative pathways involving induced homolysis of the carbon–sulphur bond to give a thioacyloxy radical have also been discussed [43].

3.2.1.1 Practical requirements

Consideration of the above mechanism leads to the conclusion that *the best way to carry out the experiment involves dropwise addition of the tin hydride to a solution of the thiocarbonyl derivative,* usually in benzene or toluene solution at a sufficiently elevated temperature to ensure that the carbonyl-forming β-scission reaction is efficient. The inverse addition mode is, however, reported in the original publication for secondary alcohol derivatives and can work well. As expected the formation of primary alkyl radicals R· [44] requires a higher operating temperature (130–140°C) while secondary alcohol derivatives usually proceed smoothly at 80–110°C. Reactions in this section have accordingly been grouped together on the basis of these practical requirements as a function of the structure of the initial alcohol.

In terms of initiator requirement, it has been found that deoxygenations involving the phenoxythiocarbonyl group [39] (X = OPh) proceed most efficiently with the addition of 0.2 equiv. of AIBN, thus implying that the overall chain length in the propagation sequence is less efficient. Initiators are not usually necessary, however, for substrates possessing xanthate, thiocarbonyl imidazolide or thionobenzoate functionality in secondary alcohols.

It is also appropriate at the outset to indicate the most commonly used conditions for preparation of the various thiocarbonyl derivatives in order that ancillary functional groups in a given substrate may be suitably protected if necessary.

(a)   Xanthate esters (X = SMe) [37]

This traditional functional group is generated under basic conditions and requires sequential reaction of the derived alkoxide anion with carbon disulphide and methyl iodide. Sodium hydride in the presence of imidazole as a catalyst is normally satisfactory. Phase transfer conditions have also been employed with success.

(b)   Thiocarbonyl imidazolide derivatives (X = 1-imidazolyl) [37]

The direct combination of the alcohol with thiocarbonyl diimidazole occurs under mild conditions, e.g. under reflux in dichloromethane, and is tolerant of a wide variety of sensitive functional groups such as epoxides which could be attacked under the more basic conditions of xanthate formation. From personal experience, we strongly recommend the use of the highly crystalline reagent prepared by the method of Fox *et al.* [37a]. The material obtained on work up is normally used directly for the radical reaction or is treated with methanol to give the readily chromatographed mixed thionocarbonate derivative (X = OMe).

(c)   Thionobenzoate esters (X = Ph) [37]

In contrast to the basic nature of the above methods, conversion to the thionobenzoate requires reaction of the alcohol with the Vilsmeier reagent prepared from *N,N*-dimethylbenzamide and phosgene, followed by passage of hydrogen sulphide with liberation of dimethylamine from the tetrahedral intermediate. Although the thionobenzoates are readily purified and stored, this method is less commonly used in polyfunctional molecules because of the strongly electrophilic character of the reagent.

(d)   Phenoxythiocarbonyl derivatives (X = OPh) [39]

These mixed thionocarbonates are relatively simple to prepare by direct reaction of the alcohol with *O*-phenylchlorothiocarbonate and 4-dimethyl-aminopyridine in acetonitrile solution. In this case, exclusive production of the alkyl radical R˙ is guaranteed by virtue of the higher energy requirement of phenyl radical production by cleavage of the aryl–oxygen bond.

3.2.2   Deoxygenation of primary alcohols via thiocarbonyl derivatives

Although, as stated, higher temperatures are required in order to ensure β-scission, this reaction can nevertheless be carried out without competing Chugaev elimination, even though addition of the stannane to the substrate is required. A systematic study [44] revealed that formation of the primary radical occurs at a lower temperature with the thiocarbonyl imidazolide and thionobenzoate derivatives than with the xanthate ester. The use of the *less* reactive tri-n-butylstannane is also preferred over triphenylstannane to avoid interception of the carbon centred radical prior to carbonyl forming fragmentation, and slow addition of the stannane to the substrate is now mandatory.

n-Octadecane from the thionobenzoate derivative of n-octadecanol

Tri-n-butyltin hydride (700 mg, 2.41 mmol) in xylene (3 ml) was added over a 2 h period to the thionobenzoate ester of octadecanol (200 mg, 0.51 mmol) in refluxing xylene (3 ml) under an argon atmosphere, and heating was continued for a further 9 h. The solvent was removed *in vacuo* and carbon tetrachloride was added to the residue with ice cooling (*CAUTION*). The solution was then heated to reflux for 3 h, cooled and the solvent removed *in vacuo*. A solution of iodine in ether was added until the iodine colour remained; the solution was then diluted with ether (70 ml) and shaken with 10% aqueous potassium fluoride (5 ml). Filtration, drying and evaporation gave a crude product which was purified by short column silica gel chromatography using pentane as the eluant to give n-octadecane (109 mg, 84%), m.p. 28°C.

From ref. [44] with permission.

Further examples are given in Table 3.4. The cases of the methyl ester of hederagenin and erythrodiol highlight that the sequence is especially useful in the case of highly hindered neopentyllic systems where traditional ionic displacement and reduction may pose problems. The latter two cases also demonstrate the possibilities for selective deoxygenation either through preferential formation of the thiocarbonyl derivative or by controlling the temperature of the stannane reduction to achieve prior deoxygenation of the secondary alcohol derivative. The yield in the carbohydrate example [45] could certainly have been improved by operating at a higher temperature, but it is nevertheless an interesting example of the β-oxygen atom effect in which radical reactions appear to work more efficiently in such an environment.

3.2.3 Deoxygenation of secondary alcohol thionocarbonyl derivatives

Historically, the application of the stannane induced deoxygenation of thionocarbonyl derivatives was first applied by Barton and McCombie [37] to secondary alcohol derivatives in their seminal paper of 1975. Since that time, numerous applications of the method have appeared from research groups throughout the world, providing appropriate testimony to its power.

The generality of the overall sequence is evident from casual inspection of the examples in Table 3.5, which represent only a small selection of the

TABLE 3.4

Deoxygenation of primary alcohol thiocarbonyl derivatives

Substrate (Z = OH)	Derivative	Solvent	Temper-ature (°C)	Product (Z = H) yield (%)	Ref.
$CH_3(CH_2)_{16}CH_2Z$	Z = O—CS—SMe	p-Cymene	150	71	[44]
	Z = O—CS—N (imidazole)	Xylene	130	81	[44]
	Z = O—CS—Ph	Xylene	130	84	[44]
Hederagenin methyl ester	Z = O—CS—SMe	p-Cymene	150	65	[44]
Erythrodiol	Z = O—CS—N (imidazole)	Xylene	130	40	[44]
	Z = O—CS—N (imidazole)			31	[45]
	Z = O—CS—SMe	Xylene	120	80	[46]

available work in the steroid, carbohydrate, terpenoid and nucleoside areas. Particularly striking examples include deoxygenation of the hindered terpene alcohol [47] where available literature methods were unsuccessful, and the

TABLE 3.5

Deoxygenation of some secondary alcohol thiocarbonyl derivatives

Substrate (Z=OH)	Thiocarbonyl derivative	Conditions	Product (Z=H) yield (%)	Ref.
	Z=O—CS—SMe	Toluene, 110°C	73	[37]
	Z=O—CS—N⟨N⟩		74	[37]
	Z=O—CS—OPh	AIBN	85	[37]
	Z=O—CS—SMe	Xylene, 140°C	83	[37]
	Z=O—CS—SMe	Toluene, 110°C	67	[37]
	Z=O—CS—SMe	Xylene, AIBN, 140°C	86	[51]
	Z=O—CS—OPh	ⁿBu₃SnH	65	[52]

TABLE 3.5 *Continued*

Substrate (Z = OH)	Thiocarbonyl derivative	Conditions	Product (Z = H) yield (%)	Ref.
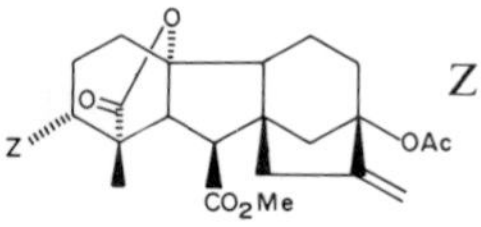	Z=O—CS—OPh		85	[53]
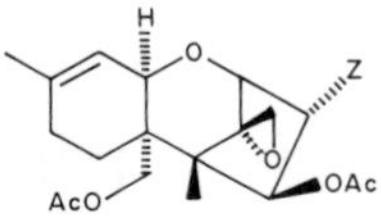	Z=O—CS—SMe		85	[54]
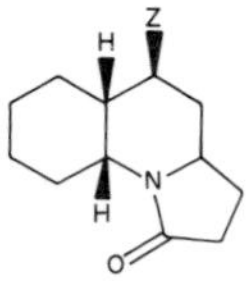	Z=O—CS—SMe		75	[55]
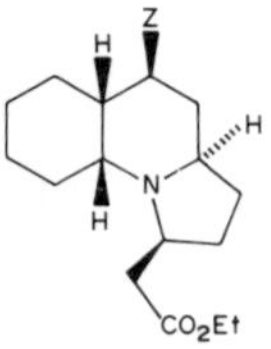	Z=O—CS—SMe		90	[55]
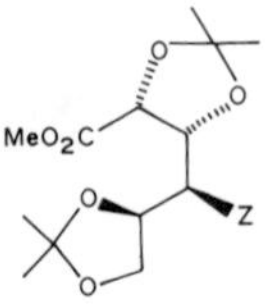	Z=O—CS—SMe		70 (overall from alcohol)	[56]
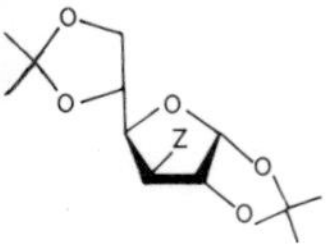	Z=O—CS—SMe	Xylene	90	[37]
	Z=O—CS—N⟨N⟩ 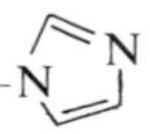	Toluene	68	[37]
	Z=O—CS—OPh	Toluene	85	[37]
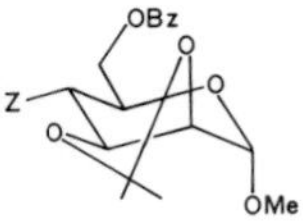	Z=O—CS—N⟨N⟩ 		87	[57]

TABLE 3.5 *Continued*

TABLE 3.5 *Continued*

Substrate (Z = OH)	Thiocarbonyl derivative	Conditions	Product (Z = H) yield (%)	Ref.
	Z = O—CS—SMe	Toluene, 110°C	94	[37]
	Z = O—CS—Ph	Toluene, 110°C	70	[37]
	Z = O—CS—SMe	Xylene, 120°C	80–84	[46]
	Z = O—CS—SMe	Xylene, 120°C	80–84	[46]
	Z = O—CS—SMe		95 (dideoxy derivative)	[58]
	Z = O—CS—SMe		78	[59]

TABLE 3.5 *Continued*

Substrate (Z = OH)	Thiocarbonyl derivative	Conditions	Product (Z = H) yield (%)	Ref.
(structure: aminoglycoside derivative with NHCO$_2$Et, AcO, EtO$_2$CHN, OAc, OBz groups)	Z = O—CS—Ph		82	[59]
(structure: tetraisopropyldisiloxane-bridged ribonucleoside, B = OMe)	Z = O—CS—OPh	AIBN	58	[39]
B = adenine (6-aminopurine, N-methyl)	Z = O—CS—OPh	AIBN	78	[39]
B = uracil/cytosine derivative (OH, O, N-methyl)	Z = O—CS—OPh	AIBN	68	[39]
B = 4-amino-indole-3-carboxamide (CONH$_2$, N-methyl)	Z = O—CS—OPh	AIBN	65	[39]
(structure: 2'-deoxyadenosine derivative, RO—, Z)	Z = O—CS—OMe	(^{n}Bu$_3$Sn)$_2$O-(MeSiHO)$_n$	88	[40]

case of the aminoglycoside antibiotic seldomycin Factor 5 [48], which was performed on a 12.5 g scale.

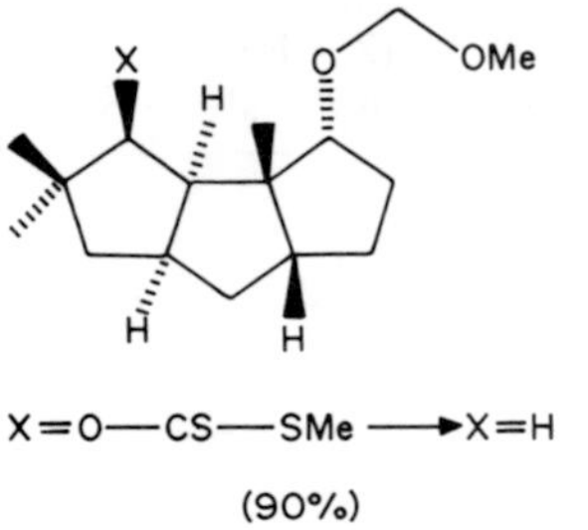

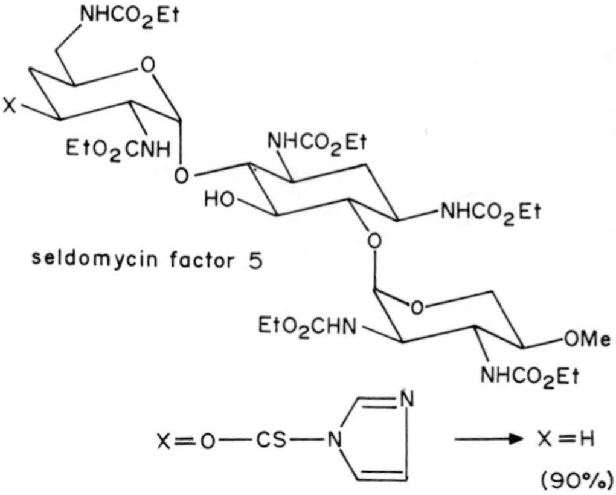

X=O—CS—SMe ⟶ X=H

(90%)

seldomycin factor 5

X=O—CS—N ⟶ X=H

(90%)

The use of tri-n-butyltin deuteride to give specifically labelled deoxy sugars is also possible. The degree of stereoselectivity is as always, however, governed by substrate geometry, but may be high in particular cases as shown [49].

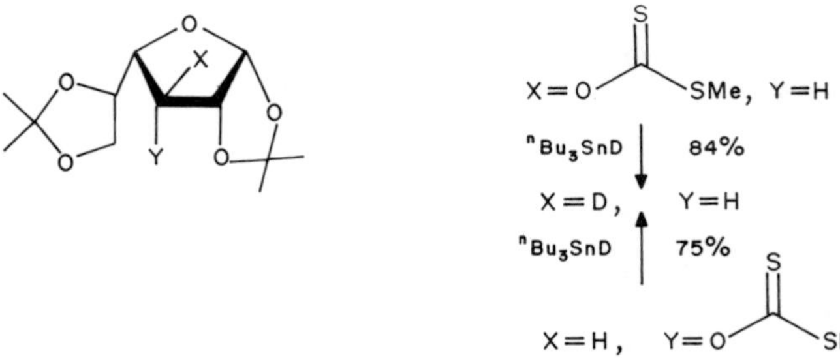

X=O $\overset{S}{\underset{}{\parallel}}$ SMe, Y=H

^{n}Bu$_3$SnD ↓ 84%

X=D, Y=H

^{n}Bu$_3$SnD ↑ 75%

X=H, Y=O $\overset{S}{\underset{}{\parallel}}$ SMe

Over the years, a number of useful additional modifications which retain the basic concept have been introduced. These include the use of readily prepared phenoxythiocarbonyl derivatives [39] (*vide supra*), which have been shown to be particularly promising for the preparation of 2-deoxy derivatives of nucleosides.

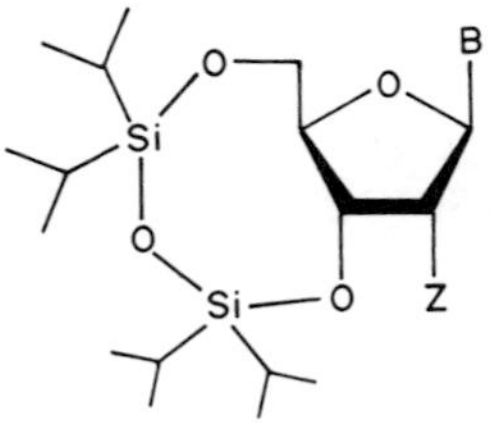

Z = O—CS—OPh Z = H   (78%)
B = purine or pyrimidine base

In a similar vein, useful practical modifications capable of wider applicability include the use of the mixed thionocarbonate [40] derived from the corresponding imidazolide by treatment with methanol; and the *in situ* generation of tri-n-butyltin hydride from bis(tri-n-butyltin) oxide and polymethylhydrosiloxane, as originally applied to the preparation of 2,3-dideoxynucleosides [40].

Deoxygenation of 3β-thiobenzyloxy-5α-cholestane—a typical procedure [37]

O-Cholestan-3β-yl thiobenzoate (510 mg) in toluene (25 ml) was added over 0.5 h to a solution of tri-n-butylstannane (450 mg) in toluene (20 ml) with refluxing under argon. After refluxing till colourless (1.5 h) the solvent was removed *in vacuo* and the residue chromatographed on alumina (grade 1), with light petroleum (b.p. 60–80°C) as eluant. Evaporation of the eluates gave cholestane (from acetone–methanol) (270 mg, 73%), m.p. and mixed m.p. 78.5–79.5°C.

A checked and highly detailed procedure in the *Organic Synthesis* series is also available [50].

3.2.4   Tertiary alcohol deoxygenation

The radical deoxygenation of a tertiary alcohol derivative possesses the advantage that the absolute stereochemistry of a neighbouring carbon centre is not destroyed, as in a sequence involving dehydration followed by catalytic hydrogenation.

The problem in this area does not lie in production of the energetically favoured tertiary radical via a β-scission reaction, but in the development of a simple and high yielding preparation of a thermally stable derivative. Xanthates, for example, are unacceptable because of the ease of Chugaev elimination. To date, two solutions based on thiocarbonyl derivatives have been found.

The first of these followed on from the observation that thionoformate derivatives of tertiary alcohols were sufficiently robust substrates [60]. A suitable route to these derivatives involves copper catalysed addition of the alcohol to *p*-dimethylaminophenyl isocyanide followed by introduction of the thiocarbonyl group using hydrogen sulphide.

Standard deoxygenation using tri-n-butyltin hydride may then be performed.

Later work, based on the development of acyl thiohydroxamates as radical precursors from carboxylic acids (Section 3.7), led to an experimentally more convenient strategy featuring deoxygenation of oxalyl thiohydroxamates [61]. These mixed oxalate esters are readily prepared by reaction of the sodium salt of 2-mercaptopyridine-*N*-oxide with the half oxalic acid chloride of the tertiary alcohol.

On reaction with a non-nucleophilic mercaptan, these derivatives undergo smooth fragmentation with loss of two molecules of carbon dioxide. Capture of the resultant tertiary radical by hydrogen atom abstraction from the thiol generates the required product and liberates the chain carrying thiyl radical. The overall yields obtained in this process are very good, as illustrated by the examples in Table 3.6.

From the experimental standpoint the key derivative is conveniently generated *in situ* by dropwise addition of the half-oxalic acid chloride to a hot benzene solution of the mercaptan containing 4-dimethylaminopyridine as catalyst and a suspension of the sodium salt of 2-mercaptopyridine-*N*-oxide. Further refinements include the use of the even less nucleophilic 3-ethylpentan-3-thiol which minimizes competitive ionic attack on the half-ester acid chloride, and the conversion of the starting alcohol into its

TABLE 3.6

Deoxygenation of tertiary alcohol oxalyl thiohydroxamates

$$(Z = O - CO - CO - O - N \diagdown)\, [61]$$

Alcohol or trimethylsiloxy derivative	Oxalyl thihydroxamate formation time (h)	Deoxygenation at 80°C		Product (Z = H) yield (%)
		Mercaptan	Time (h)	
Z=OH Z = OSiMe₃	18	A B	1.5 1.0	55 70
Z = 3α- or 3β-OSiMe₃	20	B	0.75	79 (Z = 3α-H)
	18	A	1.0	90

Mercaptans: A, 1,1-dimethylethanethiol; B, 3-ethyl-3-pentanethiol.

trimethylsilyl ether derivative for use either with alcohols which are readily dehydrated or in the presence of other acid sensitive functional groups.

General procedure for the deoxygenation of alcohols via in situ formation of oxalyl thiohydroxamates [61]

The substrate (1 mmol) in benzene (1 ml) was added at room temperature, under nitrogen to a stirred solution of oxalyl chloride (0.5 ml) in benzene (5 ml). After the reaction had been stirred for the appropriate time the solvent and excess oxalyl chloride were removed under reduced pressure. The residue was taken up in the appropriate solvent (5 ml) and added over 10 min to a stirred suspension of the sodium salt of 2-mercaptopyridine-*N*-oxide (180 mg, 1.2 mmol), DMAP (12 mg, 0.1 mmol) and a thiol at reflux under nitrogen, in the appropriate solvent (5 ml). After completion (TLC control) the cooled reaction mixture was filtered on Celite and evaporated to dryness. Chromatography on silica gel gave the pure reaction products.

3.2.5 Regioselective deoxygenation of 1,2- and 1,3-diol cyclic thionocarbonates

In an ingenious extension of the xanthate based deoxygenation protocol, Barton and Subramanıan [62] examined the reaction of tri-n-butyltin hydride with thionocarbonate derivatives, which are directly prepared by reaction of diols with *N,N*-thiocarbonyldiimidazole. As shown for the case of the D-glucose derivative, regioselectivity in the ring opening step is governed by formation of the lower energy, secondary alkyl radical, with resultant formation of 5-deoxy-1,2-*O*-isopropylidene-3-*O*-methyl-D-*xylo*hexofuranose.

Examples of both vicinal and 1,3-diol systems have been described in the carbohydrate and nucleoside areas (Table 3.7). As expected, derivatives in which both hydroxyl groups are secondary afford regioisomeric mixtures of deoxy products. However, some exceptions to this regioselectivity are known [62a].

Unlike the deoxygenation of simple secondary thiocarbonyl derivatives, the reaction is less efficient in terms of chain propagation, requiring further additions of tri-n-butyltin hydride and AIBN as initiator.

TABLE 3.7

Deoxygenation of 1,2- and 1,3-diols via thionocarbonate reduction with tri-n-butyltin hydride

Thionocarbonate	Product (yield, %)		Ref.
	(61)		[62]
	(55)		[62]
	(30) +	(60)	[62]
	(29) +	(60)	[62]
	(65)		[63]

Preparation of 5-deoxy-1,2-O-isopropylidene-3-O-methyl-
D-xylohexofuranose [62]

The thiocarbonate (0.276 g), tributyltin hydride (0.583 g), and α,α-azoiso-
butyronitrile (0.015 g) in dry toluene (15 ml) were added dropwise to refluxing
toluene (20 ml) under argon, during 45 min. Subsequent additions of the tin
hydride (2×0.292 g) together with the radical initiator (2×0.01 g) after 2 and
4 h was necessary. The reaction was complete in 6 h. The solution was treated
with aqueous sodium hydroxide 10%; 10 ml) at about 40°C for 12 h. The
organic layer was separated and the aqueous layer extracted with ether.
The combined organic extract was washed repeatedly with water until free
of base and dried (Na_2SO_4). Concentration to a syrup, followed by chromatographic elution through silica gel with light petroleum–ether mixtures of
increasing polarity, gave the title compound (0.124 g, 57%) as an oil,
$[\alpha]_D^{22} - 49°$ ($c = 2.1$).

From ref. [62] with permission.

The foregoing section has dealt exclusively with deoxygenation systems
based on the thiocarbonyl to carbonyl interconversion. The recognition of
the utility of carbon centred radical intermediates in polar molecules has also
encouraged the development of a variety of alternative methods for their
production from alcohol derivatives. Although less general in overall scope,
some of the techniques evolved may be useful in particular situations.

3.2.6 Deoxygenation of *p*-toluenesulphonate esters with tri-n-butylstannane via *in situ* iodide formation

The relatively high temperature requirements for stannane induced deoxygenation of thiocarbonyl derivatives of primary alcohols has prompted the
simple but effective one-pot method outlined below of *in situ* conversion of
a tosylate to its corresponding iodide and dehalogenated derivative [64].

Primary alcohols can be reduced by this method in excellent overall yields,
although the examples cited have not involved any problematic nucleophilic
displacements. The procedure is also applicable to secondary alcohols in useful
yields (Table 3.8).

TABLE 3.8

Deoxygenation of p-toluenesulphonate esters via *in situ* iodide displacement [64]

p-Toluenesulphonate ester	Reaction time (h)	Product	Yield (GLC) (%)
$CH_3(CH_2)_8CH_2OSO_2Ar$	1.0	$CH_3(CH_2)_8CH_3$	73[a]
$PhCH_2CH_2OSO_2Ar$	0.5	$PhCH_2CH_3$	93
$PhCH_2CH_2CH_2OSO_2Ar$	0.5	$PhCH_2CH_2CH_3$	99

	5.0		64[a]
	4.0		56
			20

[a]Isolated yield.

General procedure for the deoxygenation of p-toluenesulphonate esters [64]

Tri-n-butyltin hydride (0.24 g, 0.83 mol) was added dropwise to a refluxing solution of tosylate (0.62 mmol), sodium iodide (0.2 g, 1.33 mmol) and a catalytic amount of AIBN in 1,2-dimethoxyethane (5 ml). The mixture was refluxed for the indicated time. A usual work up gave the crude product, which was purified by column chromatography on silica gel eluted with hexane followed by chloroform to give pure product.

3.2.7　Deoxygenation of carboxylic acid esters

Although deoxygenation of simple carboxylic acid esters has been usefully achieved via electron transfer processes such as dissolving metal reduction or photolysis in aqueous hexamethylphosphoric triamide [36], their partici- pation in simple free radical chain reactions is of more recent vintage. The acetates of primary, secondary and tertiary alcohols have, however, been shown to react with 1,4-bis(diphenylsilyl)benzene in the presence of di-t-butyl peroxide as the initiator at around 140°C [65]. Several examples are collected in Table 3.9. The application of thiol catalysis to such systems would be of obvious interest for those substrates which do not contain an additional double bond.

Preparation of cyclododecane—a typical procedure [65]

A mixture of p-bis(diphenylhydrosilyl)benzene (880 mg, 2.0 mmol), cyclododecyl acetate (300 mg, 1.3 mmol) and di-t-butyl peroxide (194 mg, 1.3 mmol) was sealed in a glass tube under reduced pressure. After heating at 140°C for 15 h, cyclododecane was isolated from the reaction mixture by silica gel column chromatography using hexane as the eluant (yield, 89%).

TABLE 3.9

Deoxygenation of acetate esters by 1,4-bis(diphenylsilyl) benzene [65]

Ester	Product	Yield (%)
$CH_3(CH_2)_{10}CH_2OCOCH_3$	$CH_3(CH_2)_{10}CH_3$	95
		89
		59
		64[a]
		85

[a] 3.0 mol of silane and 2.0 mol of peroxide were used.

3.2.7.1 Deoxygenation of carbohydrate α-acetyl tertiary benzoates

Under normal circumstances benzoate esters of alcohols are totally inert to reduction by tri-n-butyltin hydride. A highly specific exception may, however, be found in the case of several α-acetyl branched tertiary benzoates of carbohydrates [66] (Table 3.10). The effectiveness of this particular reaction resides in a combination of several factors, viz. the formation of a tertiary radical which is also adjacent to a carbonyl group and the presence of additional β-oxygen atoms in the carbohydrate. Approach of the stannane to the intermediate radical determines the stereochemical outcome of the reaction which, in suitably biased substrates, can proceed with clean inversion.

TABLE 3.10

Deoxygenation of carbohydrate α-acetyl tertiary benzoates [66]

Substrate	Product	Yield (%)
		80
		(A:B = 1:4) 80

Preparation of 3-C-acetyl-3-deoxy-1,2;5,6-di-O-isopropylidene-α-D-allofuranose

The tertiary benzoate (0.20 g, 0.5 mmol) and tri-n-butyltin hydride (1.5 g, 1.5 mmol) were dissolved in toluene (25 ml) and AIBN (100 mg) added. The mixture was then brought to reflux under a nitrogen atmosphere for 2 h by means of an oil bath maintained at approximately 140°C. After cooling, silica gel (2–3 g) was added (*CAUTION*) and the toluene removed *in vacuo*. The so-formed adsorbate was slurried in hexane and deposited on a silica gel column (approximately 30 g of Kieselgel 60). The column was flushed with hexane until only pure solvent was eluted. The product was then eluted with hexane:ether (1:1). Crystallization from hexane at $-20°C$ gave the title compound as fine needles (0.11 g, 80%), m.p. 84°C, $[\alpha]_D^{20} + 50.1°$ ($c = 1$, $CHCl_3$).

From ref. [66] with permission.

3.2.8 Selective deoxygenation of tertiary and allylic alcohols via mixed methyl oxalate esters [67]

Mixed oxalate esters of secondary allylic and tertiary alcohols are readily prepared by reaction with oxalyl chloride in dichloromethane and, subsequently, methanol at room temperature. These derivatives react with tri-n-butylstannane in the presence of AIBN as the initiator, generally in refluxing toluene. The ease of deoxygenation, as monitored by the yield of isolated products, is a direct reflection of the stability of the generated alkyl radical, as illustrated by the examples from the giberellin area (Table 3.11). Selective deoxygenation is possible since the esters from primary alcohols do not suffer alkyl oxygen bond homolysis.

3.2.9 Deoxygenation reactions based on chloroformate and selenocarbonate derivatives

$$X = Cl, SePh \qquad \text{for } M = R_3Si \text{ and } R_3Sn \text{ respectively}$$

In addition to the strategies outlined above, it is also appropriate to comment on studies carried out using chloroformate and phenylselenocarbonate derivatives.

Since the silane reduction of chloroformates requires induced homolysis

TABLE 3.11

Deoxygenation of mixed methyl oxalate esters (Z = O—CO—CO—OMe)

Substrate (Z = OH)		Conditions	Product (Z = H) yield (%)	Ref.
		$^nBu_3SnH$, AIBN, toluene, 114°C	65	[67]
	X = H	$^nBu_3SnH$, AIBN, toluene	—	[67]
	X = OAc	$^nBu_3SnH$, AIBN toluene	—	[67]
		$^nBu_3SnH$, AIBN, toluene, 110°C	65	[68]

of both a relatively strong carbon–chlorine and a silicon–hydrogen bond for chain propagation, it is not surprising that a relatively large excess of di-t-butyl peroxide initiator is required, together with elevated temperatures [69]. Consequently, this method has not seen wide application in complex systems, particularly on a large scale. The conceptually similar fragmentation reaction of phenylselenocarbonates with an organostannane has also been developed by Graf [70]. Here also, however, best yields of alkane are ensured by conducting the reaction in boiling xylene or mesitylene. In this way decarboxylation of the alkoxycarbonyl radical (ROCO˙) is assured and competing hydrogen atom capture to give formate esters is minimized. It should also be noted that although selenocarbonates can be prepared in good yields from primary or secondary alcohols, yields are predictably poorer for tertiary alcohol substrates.

Deoxygenation of phenylselenocarbonates—a typical procedure [70]

A solution of the selenocarbonate (0.2 mmol) in 8 ml of solvent (*o*-xylene or mesitylene) was heated under reflux after which tri-n-butyltin hydride (1.5

mol. equiv.) and a trace of AIBN were added. Heating was continued until all of the starting material was consumed as shown by TLC control. The mixture was then concentrated and the products separated by chromatography.

Some representative examples of deoxygenation are shown in Table 3.12.

TABLE 3.12

Deoxygenation of selenocarbonates with tri-n-butyltin hydride [70]

Substrate (Z = O—CO—SePh)	Solvent	Temperature (°C)	Product (Z = H) yield (%)
	Mesitylene	164	66
	Xylene	144	54
	Xylene	144	90

3.3 DEAMINATION REACTIONS ($RNH_2 \rightarrow RH$)

3.3.1 Deamination via isocyanides, isothiocyanates and isoselenocyanates

Direct homolytic cleavage of the carbon–nitrogen bond of a primary amine is energetically unfavourable and prior transformation to a functionalized derivative is therefore required.

The elegant work of Saegusa [71] established that the isocyano group fulfils these requirements and that a simple chain reaction may be set up.

Propagation sequence

$$R - \overset{+}{N} \equiv \overset{-}{C} \xrightarrow{\text{}^n Bu_3 Sn^{\cdot}} R^{\cdot} + \text{"}^n Bu_3 SnCN\text{"}$$

$$\downarrow \text{}^n Bu_3 SnH$$

$$\text{}^n Bu_3 Sn^{\cdot} \quad + \qquad R-H$$

These derivatives are readily accessible by a two step sequence involving dehydration of the corresponding formamide. A variety of mild reagents and reaction conditions are available and permit application to sensitive polyfunctional molecules [72].

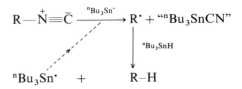

From the initial study, it was evident that cleavage of the carbon–nitrogen bond was a relatively sluggish process in comparison to a typical dehalogenation reaction, and reported yields, even for isonitriles containing secondary and tertiary alkyl groups, were modest. However, a detailed study [72] established the practical utility of the sequence. Moreover, related congeners such as isothiocyanates or isoselenocyanates, which function equally well in most cases, were shown to react through the intermediacy of the corresponding isonitrile.

$$R - N = C = S$$
$$\text{or} \qquad \xrightarrow{\text{}^n Bu_3 SnH} R - \overset{+}{N} \equiv \overset{-}{C} \xrightarrow{\text{}^n Bu_3 SnH} RH$$
$$R - N = C = Se$$

From the practical standpoint, the ease of reduction follows the expected trend, *viz.*

$$RCH_2 NC < R_2 CHNC < R_3 CNC.$$

In fact, the vigorous conditions for deamination of a simple primary isonitrile demonstrate most effectively the strength of the carbon–nitrogen bond and the amount of initiator required suggests that a chain reaction, if involved, is very inefficient. The even more demanding case of an aromatic isocyanide proved to be resistant.

Aromatic solvents have generally been employed in the presence of AIBN as the initiator. Examination of Table 3.13 shows that the procedure is compatible with free hydroxyl groups, as well as amine, sulphide, ester, amide, ketal and mesylate functionalities.

Within the domain of carbohydrate chemistry, the efficient preparation of either anomer of 2-deoxy glucose from glucosamine provides a useful practical alternative to existing methods [72].

α or β

A detailed study of selective deamination within the aminoglycoside area also revealed some features of interest. Sequential dehydration of the protected neamine derivative [73] could be controlled as indicated to provide a series of mono-, di-, tri- and tetraisonitriles.

All of these derivatives could be smoothly reduced to the corresponding desamino compounds. Moreover, as anticipated, the primary isocyano function was less reactive than the others, and could be removed or retained simply by controlling the reaction temperature.

A similar study was then carried out on gentamycin derivatives [74]. Comparison of the reaction conditions for deamination of n-octadecyl isocyanide (xylene, 140°C, 7 h, > 1.0 mol equiv. AIBN) with those of the primary isonitrile embedded in the neamine framework (benzene, 80°C, 2 h, 5 mol% AIBN) reveals an apparent anomaly.

This is, however, a specific example of a more general phenomenon, the β-oxygen atom effect [45], in which the rate of a tri-n-butylstannane reduction

TABLE 3.13

Deamination of isocyanides, isothiocyanates and isoselenocyanates with tri-n-butyltin hydride

Amine (Z = NH$_2$)	Derivative	Solvent	Temperature (°C)	Product (Z = H) yield (%)	Ref.
CH$_3$(CH$_2$)$_{16}$CH$_2$Z	Z = NC	Xylene	140	81	[72]
(steroid structure)	Z = NC	Benzene	80	89	[72]
	Z = N=C=S	Benzene	80	90	[72]
	Z = N=C=Se	Benzene	80	55	[72]
CH$_3$(CH$_2$)$_{16}$—C(CH$_3$)$_2$—Z	Z = NC	Benzene	50	90	[72]
(sugar structure, OH / OMe)	Z = NC	Toluene	114	92	[72]
(sugar structure, MeSO$_2$O / OMe)	Z = NC	Toluene	114	77	[72]
(tryptophan structure, CO$_2$Me)	Z = NC	Benzene	80	72	[77]
(amide structure, CO$_2$Et)	Z = NC	Benzene	80	71	[77]
(β-lactam structure, CO$_2$CH$_2$Ph)	Z = NC, R = CH$_2$Ph	Benzene		77	[75]
	Z = NC, R = SCH$_3$	Benzene		33	[75]

may be markedly accelerated by the presence of β-oxygen atoms in the neighbouring environment. A convincing explanation and delineation of this effect has yet to be provided, even though it does not seem to be as general as the retarding effect of β-oxygen atoms in ionic displacement reactions.

Elegant studies in the β-lactam area show that 6α-alkyl-6β-isocyano penicillin derivatives undergo highly stereoselective reduction to the 6β-alkyl derivatives, with approach of the stannane to the intermediate carbon centred radical from the more accessible convex face [75].

Penicillin chemistry also revealed that the isothiocyanate cannot be automatically regarded as an isocyanide precursor. Thus, removal of sulphur from the 6β-isothiocyanate was sufficiently slow to allow an alternative rearrangement to occur [76].

Further examples, illustrating a range of applications in natural product chemistry, are shown in Table 3.13.

The following experimental procedures illustrate typical derivitization methods, temperature ranges and initiator requirements.

Preparation of n-octadecane from n-octadecylamine: deamination of a primary isonitrile

1-Formamidooctadecane. To a saturated solution of octadecylamine (2 g) in ether–pentane was added acetic formic anhydride (0.8 g) dropwise at 0°C. After 1 h, pentane was added until the mixture became turbid, after which it was cooled to −20°C. The precipitated formamide was filtered off and recrystallized from methanol to yield the formamide (2.0 g, 91%), m.p. 66°C.

1-Isocyanooctadecane. To the formamide (0.712 g) in dry pyridine (30 ml) was added toluene-*p*-sulphonyl chloride (0.77 g). The mixture was stirred for 2 h at room temperature and poured onto ice–water. The precipitated solid was filtered off, washed with water and recrystallized from chloroform–methanol to yield the isocyanide (0.635 g, 95%), m.p. 35°C.

Reduction of 1-isocyanooctadecane with tri-n-butylstannane. A solution of the isocyanide (0.279 g) and AIBN (0.1 g) in dry xylene (50 ml) was added dropwise to a solution of tri-n-butylstannane (0.64 g, 2.2 mol equiv.) in refluxing xylene (50 ml), under nitrogen, over 2 h. A solution of AIBN (0.1 g) in xylene (50 ml) was slowly added over 5 h. The solvent was removed under reduced pressure and the residue dissolved in pentane. Iodine in pentane solution was added until the iodine colour persisted. The solvent was removed and the octadecane was isolated by preparative TLC (SiO$_2$; pentane) and sublimation *in vacuo* (0.205 g, 81%), m.p. 29°C, with spectral characteristics identical to those of an authentic specimen.

From ref. [72] with permission.

Preparation of 1,3,4,6-tetra-O-acetyl-2-deoxy-α-Dglucose: Deamination of 2-amino-2-deoxy-α-D-glucose

Glucosamine hydrochloride (35 g) was dissolved in water (50 ml) and sodium hydrogen carbonate (1.42 g) was added. A solution of *p*-nitrophenyl formate (5.42 g, 2 equiv.) in dioxan (50 ml) was added and the mixture was stirred for 48 h at room temperature. The dioxan was removed under reduced pressure and the residual aqueous phase was extracted with ether (2 × 50 ml) and then concentrated to dryness. Acetic anhydride (50 ml) and pyridine (5 ml) were added and the mixture was stirred at room temperature for 2 days, concentrated to dryness, and the residue dissolved in chloroform. The inorganic residues were filtered off and the filtrate was evaporated to yield the *N*-formylamino sugar as a foam in quantitative yield, $\bar{v}_{max}(CHCl_3) =$ 3415, 1740 and 1690 cm^{-1}, $\delta = 7.1$ (1 H, br s), 6.2 (1 H, d, $J = 5$ Hz), 3.5–6.0 br m) and 1.9–2.2 (12 H, overlapping acetyl signals).

Without further purification this product, dissolved in dichloromethane (200 ml) was cooled to $-30°C$ and triethylamine (15 ml) was added. Phosphoryl chloride (2.5 g) was added dropwise over 10 min. The mixture was stirred and allowed to warm to room temperature. After 2 h, TLC (SiO$_2$; toluene–tetrahydrofuran, 22:3) indicated that the reaction was incomplete. Further phosphoryl chloride (2.5 g) and triethylamine (15 ml) were then added. After 6 h no starting material could be detected and the reaction mixture was concentrated to 25 ml and loaded onto a silica column (5 × 20 cm) packed in toluene. (An exothermic reaction took place on the column but this in no way affected the separation.) Elution with toluene–tetrahydrofuran (22:3) gave the isocyanide (5.17 g, 89%) which was chromatographically homogenous, $\bar{v}_{max}(CHCl_3) = 2140$ and 1735 cm^{-1}, $\delta = 6.3$ (1 H, d, $J = 3$ Hz), 3.5–6.0 (6 H, m) and 1.9–2.3 (12 H, overlapping acetate signals).

To a stirred solution of tri-n-butylstannane (4.63 g) in dry toluene (50 ml) at 85°C, was added dropwise over 15 min the crude isocyanide and AIBN (0.1 g) in dry toluene (50 ml). After 2 h, TLC indicated that some unchanged isocyanide was still present and a further quantity of tri-n-butylstannane (1 g) and AIBN (0.1 g) was added. After a further 2 h, the infrared spectrum showed no $-N\equiv\bar{C}$ stretch and the mixture was cooled, reduced in volume, and treated with a solution of iodine in toluene until the iodine colour persisted. The solution was filtered through a short silica column and concentrated to dryness. The colourless product was dissolved in acetonitrile (25 ml) and the solution washed with pentane (2 × 25 ml), concentrated to dryness and the residual solid dissolved in ether (50 ml). The ethereal solution was washed with aqueous potassium fluoride (2 × 25 ml, 10% w/v), dried (MgSO$_4$), filtered and evaporated to yield the tetraacetate (4.35 g, 81% based on glucosamine hydrochloride), m.p. 105–107°C (literature, 106°C), $\bar{v}_{max}$ (CHCl$_3$) = 1740 cm^{-1}, $\delta = 6.2$ (1 H, m), 4.8–5.5 (2 H, m), 3.9–4.5 (3 H, m) and 1.9–2.1 (14 H, m), $[\alpha]_D^{20} + 104°$ ($c = 0.24$ in EtOH) (literature, $+117°$ (MeOH)). Although the melting point and spectroscopic properties indicated the product to be pure, it was possible to detect the smell of organotin residues. Repeated recrystallization from ethanol removed this odour and the melting point increased to 109–110°C and the specific rotatory power to 107°.

From ref. [72] with permission.

3.4 DENITRATION REACTIONS (RNO$_2$ → RH) [78]

The direct replacement of the nitro group by a hydrogen or deuterium atom from tri-n-butyltin hydride or deuteride was first reported in 1981 [79], and has already proven to be a simple and effective strategy because of the synthetic versatility of this functional group in a variety of fundamental carbon–carbon bond forming reactions.

From the mechanistic standpoint, the chain sequence has been the subject of some debate. An unusual electron transfer mechanism was originally proposed and EPR evidence was used in support of the radical anion of the nitro compound [80].

Propagation sequence

$$RNO_2 \; + \; {}^nBu_3Sn^{\cdot} \longrightarrow {}^nBu_3Sn^{+} \; + \; [RNO_2]^{\cdot -}$$

$$\downarrow {-NO_2^{-}}$$

$${}^nBu_3Sn^{\cdot} \; + \; RH \xleftarrow{\;{}^nBu_3SnH\;} R\cdot$$

An alternative study by Ono [78, 81] suggests, however, that the EPR signal present in the reaction mixture should be attributed to the nitroxide radical produced by addition of the tin radical to the oxygen atom of the nitro group.

Propagation sequence

$$R{-}NO_2 \; + \; {}^nBu_3Sn^{\cdot} \longrightarrow R{-}N(O^{\cdot}){-}O{-}SnBu_3$$

$${}^nBu_3Sn^{\cdot} \; + \; RH \xleftarrow{\;{}^nBu_3SnH\;} R\cdot \; + \; {}^nBu_3Sn{-}ONO$$

In preparative terms, the ease of reductive replacement is readily summarized as a function of substrate structure. Thus:

(i) Primary alkyl nitro compounds (RCH_2NO_2) and aromatic nitro compounds do not undergo denitration.

(ii) For simple secondary alkyl nitro groups, a five-fold molar excess of stannane and 0.5 M equiv. of an initiator such as AIBN in toluene at 110°C are required.

(iii) Tertiary alkyl nitro compounds undergo smooth denitration using 1.2 M equiv. of stannane and 0.2 M equiv. of AIBN in benzene at 80°C.

(iv) When an electronically stabilized radical is produced after splitting off the nitro group, as in the case of benzylic, allylic or β-keto nitro compounds, reduction is straightforward and reaction conditions as for tertiary nitro compounds may be applied.

In effect, the experimental procedures are generally straightforward.

Method A.   General procedure for the denitration of tertiary nitro compounds [78, 79]

A mixture of the nitro compound (10 mmol), tri-n-butyltin hydride (12 mmol) and AIBN (2 mmol) in benzene (5 ml) is heated at 80°C for 1–2 h. The pure product is obtained by column chromatography on silica gel (hexane/ethyl acetate) and/or distillation (yield, 80–95%).

Method A is also applicable to nitro compounds possessing adjacent benzylic, allylic or carbonyl functionalities.

Method B.   General procedure for the denitration of secondary nitro compounds [78, 82]

A mixture of the nitro compound (10 mmol), tri-n-butyltin hydride (50 mmol) and AIBN (5 mmol) in toluene (5 ml) is heated at 110°C for 0.5–1 h. The pure product is isolated by column chromatography on silica gel (hexane/ethyl acetate) and/or distillation (yield, 50–69%).

In terms of synthetic planning, the greatest advantage of the present reaction is that the usefulness of the nitro group, either in ionic or pericyclic reactions, may be harnessed in a previous step as an activating group or regiocontrol element for carbon–carbon bond formation, and then coupled with its ability to "vanish" on reaction with tri-n-butylstannane.

Thus consideration of the nitronate anion as a carbon nucleophile gives *inter alia*:

(1) Regiocontrolled conjugate addition of polyfunctionalized alkyl groups [83].

(2) An unsymmetrical ketone synthesis [78, 84, 85] with potential for selective deuteration [86].

(3) Controlled hydroxyalkylation [78].

(a)

(b) As applied, at the anomeric centre of carbohydrates, the reaction is predictably highly stereoselective [87].

In Diels–Alder reactions, the dominant dienophilicity of the nitro group may be used not only to further activate the dienophile, but also to control the regiochemistry in the adduct. Thus, in the example shown, high yield reductive removal with tri-n-butyltin hydride then furnishes the opposite regioisomer to that obtained *via* cyclohexenone itself [88, 89].

Nitroalkenes are also useful acceptors, as in the prostaglandin synthesis [90] involving a tandem organocopper addition to the cyclopentenone derivative and trapping of the resultant enolate. Removal of the secondary nitro group was accomplished by reduction with tri-n-butylstannane in refluxing toluene.

Finally, mention should be made of the stability of the nitronate anion acting as the thermodynamic sink for a smooth rearrangement–fragmentation approach to macrocylic lactones which does not require high dilution conditions [82]. Removal of the relatively unactivated secondary alkyl nitro group was achieved in 48% yield.

Further selected examples in Table 3.14 indicate that, in appropriate cases, highly chemoselective transformations in the presence of keto, ester, cyano and even chloride and sulphide functionalities may be carried out.

TABLE 3.14

Reduction of aliphatic nitro compounds by tri-n-butylstannane in the presence of AIBN

Derivative (Z = NO$_2$)	Method	Product (Z = H) yield (%)	Comment	Ref.
(structure: CN)	A	85		[78, 79]
(structure: SPh, CO$_2$Me)	A	75		[78, 79]
(structure: CO$_2$Et)	A	48		[78, 79]
(structure: sulfoxide, S=O, Ph)	A	93	Functional group compatibility	[78, 83]
(structure: (CH$_2$)$_7$CO$_2$Me)	A	78		[85]
CH$_3$(CH$_2$)$_6$ (structure, H)	A	87		[78, 91]
(structure, cyclohexanone)	A	90	From (structure with NO$_2$) + (enone) Catalysed by Ph$_3$P non-basic conditions	[78, 91]
OAc (structure, isopropyl)	B	45	Secondary nitro compound	[89]

TABLE 3.14 *Continued*

Derivative	Method	Product (Z = H) yield (%)	Comment	Ref.
	A	95	Dithiolane retained	[89]
	A A (^{n}Bu$_3$SnD)	67 65 (Z = D)		[86]
	A	85	Precursor prepared from + KCN $K_3Fe(CN)_6$ $R_2CHNO_2 \rightarrow R_2CHCN$ conversion	[81]
	A	78	$R_2C \rightarrow R_2CHP(O)(OR')_2$	[92]

3.5 REDUCTIVE HOMOLYTIC CLEAVAGE OF THE CARBON–SULPHUR BOND

3.5.1 Introduction

In general terms, free radical chain reactions involving induced homolysis of a simple unactivated carbon–sulphur bond are energetically demanding, relative to many of the other functional group triggers for generation of carbon centred radicals. For this reason, many reactions involving, for example, dehalogenation, deselenation or denitration can be successfully carried out in the presence of a remote sulphide functionality [93]. Nevertheless, several synthetically useful systems featuring a designed radical cleavage of a carbon–sulphur bond have been reported.

3.5.2 Thiol desulphurization (RSH → RH)

The simplest, cheapest and most effective chain reaction for removal of a
sulphur atom from a mercaptan remains the well established reaction with
a trialkyl phosphite in the presence of light or radical initiators [94, 95]. The
key step following addition of the thiyl radical to the phosphorus centre is
a highly efficient β-scission reaction to release the alkyl radical.

Propagation sequence

$$RS^{\cdot} + P(OEt)_3 \longrightarrow R\overset{\frown}{S}\overset{\frown}{P}(OEt)_3$$

β-scission

$$RS^{\cdot} + RH \overset{RSH}{\longleftarrow} R^{\cdot} + S = P(OEt)_3$$

The high yield in the desulphurization of n-octyl mercaptan, which involves
liberation of a primary alkyl radical, provides an excellent example of the
thermodynamic driving force provided by formation of the pentavalent
thionophosphate.

Preparation of n-octane from n-octyl mercaptan [94, 95]

$$^{n}C_8H_{17}SH \xrightarrow[P(OEt)_3]{h\nu} {}^{n}C_8H_{18}$$

A mixture of 83 g (0.5 mol) of triethyl phosphite (previously distilled from
sodium) and 73 g (0.5 mol) of n-octyl mercaptan in a Pyrex flask fitted with
a 24 inch (60 cm) fractionating column was irradiated with a General Electric
100 W S-4 bulb at a distance of 5 inches (~ 12.5 cm) from the flask. After
6.25 h of irradiation, the mixture was distilled to give 50.3 g (88%) of octane,
b.p. 122.0–124.5°C, n_D^{25} 1.3951–1.3959 (reported b.p. 125/59°C, n_D^{25} 1.3953)
and 90.9 g (0.459 mol) of triethyl thionophosphate, b.p. 45°C (0.50 mmHg),
n_D^{20} 1.4461 (reported, b.p. 105–106°C (20 mmHg)), n_D^{20} 1.4480; calculated n_D^{25}
1.4460).

More recent work has shown that reaction of a mercaptan with at least
two equivalents of tri-n-butyltin hydride in the presence of AIBN leads *via*
formation of an intermediate alkyltin sulphide to the corresponding
hydrocarbon [96–98].

$$RSH \xrightarrow{^{n}Bu_3SnH} RS-SnBu_3 \xrightarrow{^{n}Bu_3SnH} RH$$

The ease of a reductive desulphurization parallels the stability of the intermediate radical in the conventional way. Aromatic thiols, however, are not cleaved and lead only to the organotin thiophenate.

A particularly elegant application [97] which includes this method is the reduction of the β-acyloxymercaptan macrolide which was, in turn, prepared by intramolecular acyl transfer.

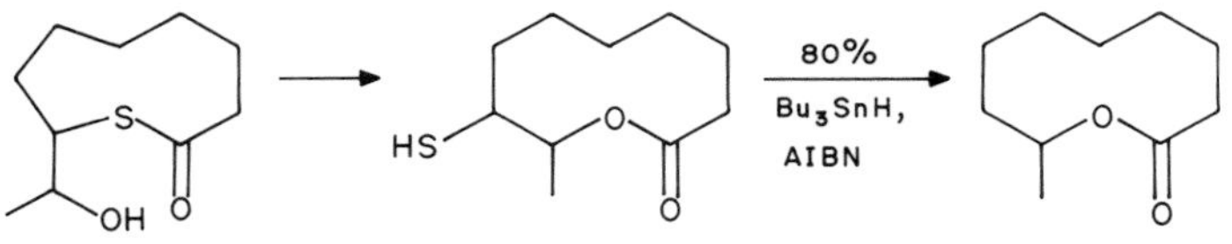

80%

Bu₃SnH, AIBN

Desulphurization of mercaptans [97, 98]

A mixture of the thiol and tri-n-butylstannane (2.1–1.2 M equiv.) with AIBN as the initiator were heated in benzene solution (0.2 M) for 14 h at 80°C. Removal of the solvent and chromatography furnished the corresponding desulphurized derivative.

The yields of some representative examples in Table 3.15 are an appropriate reflection of the energy of the intermediate radical.

TABLE 3.15

Desulphurization of mercaptans with tri-n-butylstannane, and AIBN

Mercaptan (Z = SH)	Product (Z = H) Yield (%)	Ref.
$CH_3(CH_2)_{10}CH_2Z$	65	[97]
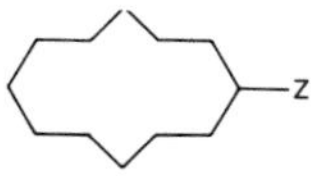 (cyclododecyl)–Z	59	[97]
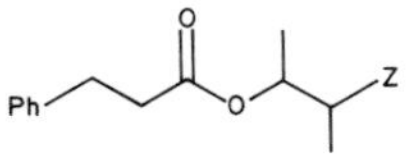 Ph....O....Z	90	[97]
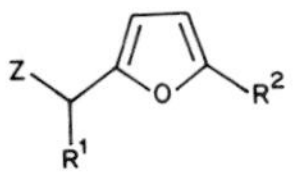 Z....O....R² (furan, R¹)		
$R^1 = R^2 = H$	44	[98]
$R^1 = COPh, R^2 = H$	66	[98]
$R^1 = CO_2Et, R^2 = Et$	89	[98]

3.5.3 Reductive cleavage of sulphides

In contrast to the parent mercaptans, the derived methyl sulphides give much lower yields [97] on attempted desulphurization with tri-n-butyltin hydride. Similarly, di-n-octyl sulphide gives n-octane in only 8% yield [99].

Synthetically useful reactions involving regioselective cleavage have accordingly featured either the use of unsymmetrical alkyl aryl sulphides, and/or cases in which the derived alkyl radical R˙ is energetically favoured. Aryl sulphides with unactivated primary or secondary alkyl groups are extremely sluggish to react [99].

$$R\!-\!S\!-\!Ph \xrightarrow{\ ^{n}Bu_3Sn˙\ } R˙ + {}^{n}Bu_3Sn\!-\!S\!-\!Ph$$

$$\Big\downarrow {}^{n}Bu_3SnH$$

$$RH$$

TABLE 3.16

Desulphurization of unsymmetrical sulphides by tri-n-butylstannane and AIBN

Substrate	Product	Yield (%)	Ref.
$CH_3(CH_2)_6SCH_2Ph$	$PhCH_3$ (85), ${}^{n}C_7H_{16}$	49	[99]
		97	[101]
	(After aminal hydrolysis and nitrogen protection)		[102]
	$\alpha:\beta = 1:12$	87	[103]
		38	[104]

The sequence has been shown to be:

R = Ph (not cleaved) ≪ CH$_3$ < primary alkyl < secondary alkyl
< tertiary alkyl < allyl < benzyl.

Carbon–sulphur bonds adjacent to carbonyl groups or heteroatoms have proven most successful, as shown by the selected examples in Table 3.16, the only exception to the above rules being the final carbohydrate example where, once again, operation of the β-oxygen atom effect (p. 60 may be in evidence.

Desulphurization of 2,3-dimethyl-7-(1-hydroxy-1-methylethyl)-7-phenylthio-8-oxo-1-azabicyclo[4.2.0]octane [100]

3.3 : 1

A mixture of the azetidinone sulphides (3.4:1) (20 g, 62 mmol), AIBN (2.0 g), tri-n-butyltin hydride (56.4 ml, 210 mmol) and acetone (500 ml) was refluxed for 16 h under nitrogen. The mixture was evaporated to dryness and the residue was dissolved in chloroform. After filtration to remove insoluble materials, the filtrate was concentrated and the residue was treated with diisopropyl ether to give the crystalline *cis*-azetidinone (8.13 g, 61.3%). The mother liquor was concentrated and the concentrate was subjected to chromatography on silica gel. Elution with hexane–ethyl acetate (1:1 v/v) gave an additional crop of the *cis*-azetidinone (1.70 g, 12.8%) and the crystalline *trans*-azetidinone (2.63 g, 19.8%).

From ref. [100] with permission.

3.5.4 Reduction of dithioacetals

The reduction of dithioacetal derivatives of aldehydes and ketones [105] with tri-n-butyltin hydride offers a complementary method to the more commonly employed Raney nickel reaction. As anticipated, the use of AIBN as the initiator is necessary, and, as required by the stoichiometry of the reaction, 2 equiv. of hydride are needed for complete reduction.

The use of 1 equiv. of stannane allows the reaction to be stopped cleanly at the intermediate sulphide stage and the mercaptan can then be liberated by silica gel hydrolysis.

Although reactions, are, as always, substrate dependent, the following procedure is typical.

Desulphurization of 1,4-dithiaspiro[4.5]decane

To a stirred solution of 2.09 g (0.012 mol) of cyclohexanone ethylene dithioketal and 13.97 g (0.048 mol) of tri-n-butyltin hydride was added 80 mg of recently recrystallized AIBN. The reaction mixture was reacted at 80°C for 1.5 h and then short-path distilled at house vacuum pressures (90–160 mmHg). The volatile material was trapped in a receiving flask cooled in a dry ice–acetone bath, yielding 0.79 g (76%) of cyclohexane which gave NMR and infrared spectra identical to an authentic sample. The pot residue

TABLE 3.17

Desulphurization of dithioacetals by tri-n-butyltin hydride and AIBN

Substrate	$^nBu_3SnH$ (mol equiv.)	Product	Yield (%)	Ref.
Ph	2.0	$PhCH_3$	73	[105]
MeO	2.0	MeO	95	[105]
$CH_3(CH_2)_5$	2.0	$CH_2(CH_2)_5CH_3$	80	[105]
	1.0		64	[105]
	2.0		65	[106]

was further distilled at reduced pressure to give 4.55 g (62%) of bis(tri-n-butyltin) sulphide (b.p. 204–206°C (20 mmHg)) which gave an infrared spectrum identical with an authentic sample.

From ref. [105] with permission.

3.5.5 Reduction of thiocarbonyl groups ($R_2CS \rightarrow R_2CH_2$) [107]

The reduction of the thione moiety provides a perfect illustration of the importance of controlling experimental conditions in free radical chain reactions. Thus, in contrast to the controlled selective monodeoxygenation of a vicinal diol (Section 3.2.5) which is achieved via slow addition of tri-n-butyltin hydride to the substrate, reductive desulphurization requires immediate capture of the first formed intermediate radical and hence a high concentration of tri-n-butyltin hydride.

A variety of usefully functionalized heterocyclic systems can be obtained in this way, as shown by the examples in Table 3.18.

TABLE 3.18

Desulphurization of thiocarbonyl groups using tri-n-butylstannane (3.0–5.0 mol equiv.) and AIBN in toluene [107]

Substrate	Product	Yield (%)
		55
		75
		75
		94

3.6 DESELENATION REACTIONS [108]

3.6.1 Introduction

The rich ionic chemistry of organoselenium compounds [108], and in particular of the phenylseleno functionalization process have been extensively employed in organic synthesis. Consequently, the high yielding reductive removal of the phenylseleno group by organostannanes undoubtedly contributes to the overall value of such sequences, offering an alternative to oxidative selenoxide elimination.

$$XH = OH, SH, CO_2H$$

The analysis previously applied to the case of unsymmetrical sulphides (Section 3.5.3) can be directly transposed to organoselenium compounds, with the result that regiospecific cleavage of alkyl aryl selenides is guaranteed to occur as shown. The more facile cleavage of the weaker carbon–selenium bond does, however, offer significant advantages in terms of milder reaction conditions over the situations obtaining in mechanistically analogous sulphur counterparts.

In practical terms, triphenyltin hydride is the best reagent [109] and reductions are generally performed by addition of the stannane to a refluxing toluene solution of the substrate. The addition of AIBN as an initiator is also recommended, leading to faster reactions at lower temperatures. Chromatographic removal of the tin selenide by-product is particularly easy in deselenation reactions since "trailing" does not occur.

3.6.2 Reductive cleavage of unsymmetrical selenides

The discovery by Clive [109] of the reductive removal of the phenylseleno group has led to the incorporation of this sequence in many synthetic strategies.

The usual rules that delivery of the hydrogen or deuterium atom will occur from the less hindered face of a sterically demanding substrate can be applied, as illustrated by the clean reduction of both isomers of the 6-phenylselenopenicillanates to the 6β-substituted reduction product [75, 110].

In contrast to sulphides, the facile cleavage of primary alkyl phenyl selenides is now a useful reaction. Since these, in turn, are readily prepared by reaction of the corresponding alcohol with a phosphine and diphenyl diselenide, the overall sequence is a useful alternative to thiocarbonyl based strategies, especially for deoxygenation of primary alcohols [109, 111].

$$RCH_2OH \xrightarrow[\text{PhSeSePh}]{R^1_3P} RCH_2SePh \xrightarrow{Ph_3SnH} RH$$

High yield differentiation has also been demonstrated in the case of vinyl methyl selenides [112].

An extensive range of examples (Table 3.19) indicates that the reaction may be performed in the presence of ether, hydroxyl, sulphide, lactone and urethane functionalities.

General procedures for the reduction of alkyl phenyl selenides

Method 1. The apparatus consisted of a small, round bottomed flask containing a magnetic stirring bar and carrying a reflux condenser closed with a septum through which were passed inlet and exit needles for nitrogen. (In order to avoid mechanical losses in very small scale experiments it is preferable to have the flask fused to the condenser so that the apparatus is a one piece unit.) The selenide (or selenoacetal) was weighed into the flask, solvent (toluene or benzene) was added from a syringe and the septum was

TABLE 3.19

Reduction of alkyl phenyl selenides by organostannanes in the presence of AIBN

Substrate (Z = SePh)	Method	Product (Z = H) yield (%)	Ref.
$CH_3(CH_2)_{10}CH_2Z$	1	73	[109]
	1	89	[109]
	1	60	[109]
	1	62	[109]
	1	72	[109]
	1	70	[109]
	2	89	[113]
	1 2	74 97	[109] [113]
	2	76	[113]

fixed in place. Nitrogen was swept through the system for about 5 min and then the exit needle was removed so that the contents of the flask were kept under a slight static pressure of nitrogen. Triphenyltin hydride was injected into the mixture and the flask was lowered partially into a preheated oil bath (120–125°C for toluene). The course of the reaction was monitored by TLC, samples for examination being withdrawn through the condenser septum. Sometimes additional portions of triphenyltin hydride were added at intervals. The product was isolated by chromatography and/or distillation.

From ref. [109] with permission.

The use of tri-n-butylstannane has also proved useful, as employed by the Nicolaou school [113].

Method 2.   *Preparation of 6-oxabicyclo[3.2.2]nonan-7-one.*

A solution of the phenylselenolactone (295 mg, 1 mmol) in freshly distilled toluene (5 ml) was mixed with tri-n-butyltin hydride (582 mg, 390 1, 2 mmol) and a 0.02 M toluene solution of AIBN (1 ml, 0.02 mmol). The mixture was degassed with a stream of argon for 15 min, sealed with a plastic cap and heated to 110°C for 1 h. Removal of the solvent and column chromatography of the residue (silica, CH_2Cl_2) afforded pure 6-oxabicyclo[3.2.2]nonan-7-one (123 mg, 88%) as an oil which crystallized on standing.

From ref. [113] with permission.

3.6.3   Reduction of selenoketals [109]

As in the case of their sulphur analogues, stannane reduction provides an alternative to the classical Wolff–Kishner reaction. Reaction conditions are relatively mild and the required derivatives are readily prepared using tris(phenylseleno)boron.

TABLE 3.20

Reduction of selenoketals by organostannanes in the presence of AIBN

Substrate	Derivative	Stannane	Product (Z = H) yield (%)	Ref.
$CH_3(CH_2)_9CH\begin{smallmatrix}Z\\Z\end{smallmatrix}$	Z = SePh	Ph$_3$SnH	84[a]	[109]
(steroid structure, AcO)	Z = SePh	Ph$_3$SnH	73	[109]
(steroid structure, MeO)	Z = SePh	Ph$_3$SnH	64	[109]
(tetralin structure)	Z = SePh	Ph$_3$SnD	89(Z = D)	[109]
(cyclohexanone structure)	Z = SeMe	^{n}Bu$_3$SnH	94	[114]

[a]GLC yield.

The range of examples in Table 3.21 provides a representative series of alkyl phenyl selenides (Section 3.6.2, method 1) and some representative examples are shown in Table 3.20.

3.6.4 Deselenation of acyl selenides [70]

Cleavage of the carbon–selenium bond in acyl selenides provides a very mild method for the generation of acyl radicals, the required phenylselenoesters being derived, in turn, by nucleophilic displacement from the corresponding acid chloride with phenylselenol.

TABLE 3.21

Tri-n-butyltin hydride reduction of acyl selenides [70]

			Product yield (%)	
Substrate (Z = COSePh)	Solvent	Temperature (°C)	Aldehyde (Z = CHO)	Nor-alkane (Z = H)
	Benzene	80	92	8
	Mesitylene	164	13	84
	Benzene	80	1	98
	Benzene (hv)	20	18	79
	Benzene	80	82	17
	Benzene (hv)	20	83	3
	Mesitylene	164	18	80
	Benzene	80	—	82
	Benzene	80	69	
	Mesitylene	164	51	

The fate of the acyl radical is a function of its structure and of the temperature at which the reaction is carried out. Thus, decarbonylation to give nor-alkanes is favoured by an increase in temperature and for substrates leading to tertiary alkyl radicals. Very good yields of aldehydes can be obtained at lower temperatures and in cases where decarbonylation is more energetically demanding, such as α,β-unsaturated acyl selenides. The reaction can also be carried out at 20°C under photochemical conditions.

General procedures for the conversion of acyl selenides to aldehydes or nor-alkanes [70]

(a) Thermal method. A solution of the selenoester (0.2 mmol) in 8 ml solvent (benzene, toluene, o-xylene or mesitylene; see Table 3.21) was heated under reflux after which tri-n-butyltin hydride (1.5 mol equiv.) and a trace of AIBN were added. Heating was continued until all the starting material was consumed (TLC control, 2–60 min). The mixture was then concentrated *in vacuo* and the products separated by chromatography.

(b) Photochemical method. A solution of the selenoester (0.2 mmol) in benzene (8 ml) and tri-n-butyltin hydride (1.5 mol equiv.) was irradiated in a quartz vessel using a low pressure mercury lamp until all the starting material was consumed (TLC control), after which it was worked up as described for the thermal method.

The range of examples in Table 3.21 provides a representative series of conditions and indicates that diene, olefin, aromatic and ester functionalities have been shown to be compatible.

3.7 DECARBOXYLATION REACTIONS [115]

3.7.1. Introduction

Throughout the development of radical chemistry, the rapid loss of carbon dioxide from a carboxyl radical has proven to be an effective approach for alkyl radical generation. Nevertheless, problems such as carbocation generation by overoxidation in the electrolytic Kolbé procedure, or the relatively inefficient nature of the non-chain thermolysis of peresters, have hindered maximum exploitation of this method.

Initial efforts by the Barton group [116] to develop a propagative chain decarboxylation sequence included the introduction of phenylthiodihydrophenanthrene esters which underwent smooth reductive decarboxylation on reaction with tri-n-butylstannane.

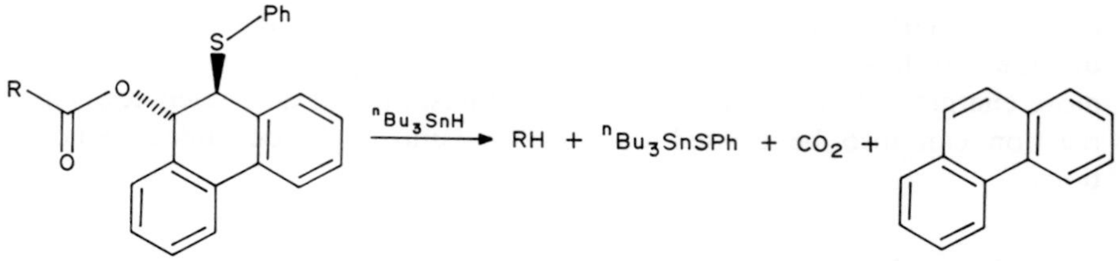
Ph
S
R
O
O
ⁿBu₃SnH
RH + ⁿBu₃SnSPh + CO₂ +

Further development of these derivatives was, however, hampered by the lack of generally efficient ionic esterification methods. Novel derivatives were accordingly sought, culminating in the introduction of the *O*-acyl derivatives of the thionohydroxamic acid *N*-hydroxypyridin-2-thione [117].

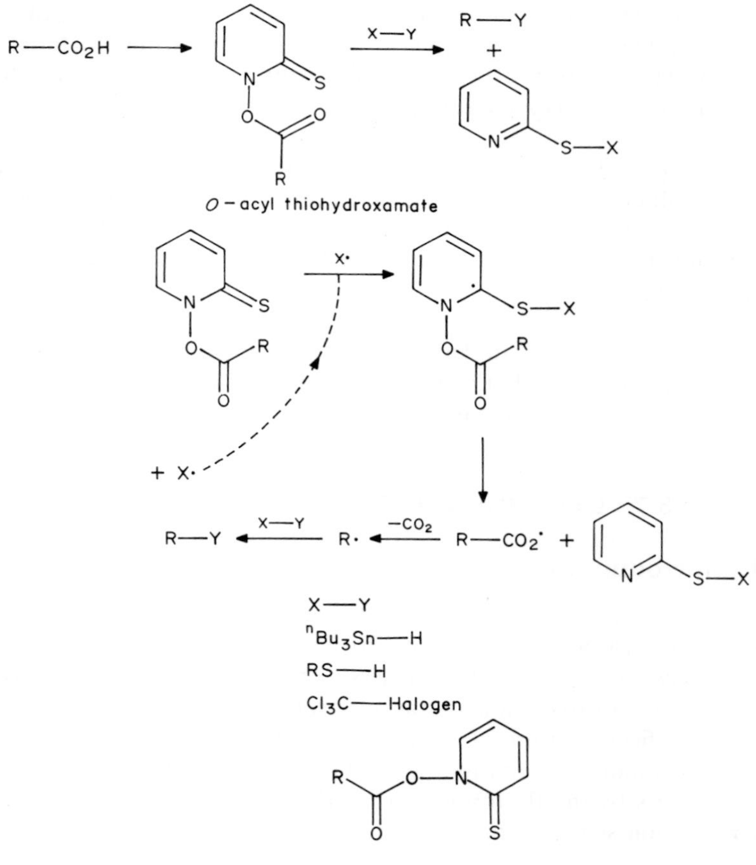
R—CO₂H
X—Y
R—Y
+
S—X
O-acyl thiohydroxamate
X·
+ X·
R—Y
X—Y
R·
−CO₂
R—CO₂˙ +
X—Y
ⁿBu₃Sn—H
RS—H
Cl₃C—Halogen

The extensive series of investigations outlined below revealed that these derivatives undergo facile reaction not only with tri-n-butylstannyl radicals as the chain carrier X•, but also with thiyl, alkyl, trichloromethyl and other thiophilic radical species in a rich and productive way. In the general chain mechanism shown, the reagent X–Y therefore serves as a source of the thiophilic radical X•, and is also capable of reacting with the resultant alkyl radical R•, in an efficient atom or group transfer reaction to give the product R–Y.

In contrast to the dihydrophenanthrene derivatives, O-acyl thiohydroxamates are readily prepared [117–119] either from the thionohydroxamic acid itself or from its commercially available sodium salt by reaction with either an acid chloride or the mixed anhydride prepared from the carboxylic acid and isobutyl chloroformate. Additional preparations include classical activation of the carboxylic acid with dicyclohexyl carbodiimide or treatment of the triethylammonium salt of the acid with an activated heterocyclic salt which is prepared in quantitative yield from the thionohydroxamic acid and phosgene. Although isolable, a further advantage of the overall methods is that the derivatives may often be conveniently generated in situ prior to their application in the free radical chain reaction.

3.7.2 Nor-alkanes by reductive decarboxylation of O-acyl thiohydroxamates ($RCO_2H \rightarrow RH$) [118]

$$X = {}^{n}Bu_3Sn, \; Me_3CS, \; {}^{n}C_{11}H_{23}Me_2CS$$

The capture of the alkyl radical R• derived from the O-acyl thio-hydroxamate by a stannane or a non-nucleophilic tertiary thiol provides an efficient route for reductive decarboxylation to the derived nor-alkanes. In terms of practical expediency in ease of work up, use of the mercaptan is

TABLE 3.22

Reductive decarboxylation of carboxylic acids via O-acyl thiohydroxamates

Substrate ($Z = CO_2H$)	Method	Solvent	Temperature (°C)	Product ($Z = H$) yield (%)	Ref.
	B	Toluene	110	62	[118]
	A	Toluene	110	46	[118]
	A	Benzene	80	77	[118]
	C	Toluene	110	77	[118]
	B	Toluene	110	82	[118]
	A	Toluene	110	79	[118]
	A	Benzene	80	65	[118]
	A	Toluene	110	57	[118]

TABLE 3.22 (*continued*)

Substrate (Z = CO$_2$H)	Method	Solvent	Temperature (°C)	Product (Z = H) yield (%)	Ref.
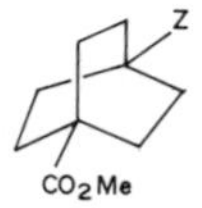	D	Benzene	80	94	[120]
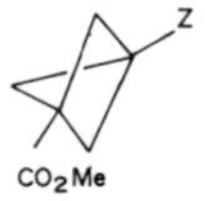	D	Benzene	80	74	[120]
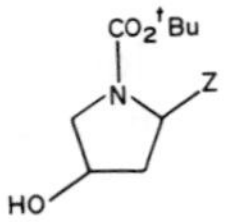	E	Tetrahydrofuran	Room temp. (*hν*)	69	[119]
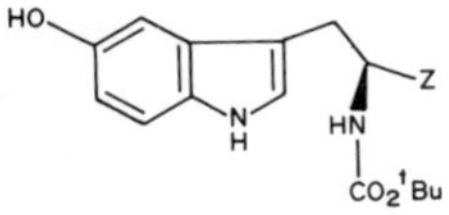	E	Tetrahydrofuran	Room temp. (*hν*)	94	[119]
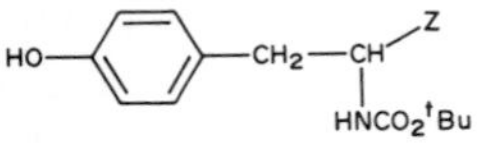	E	Tetrahydrofuran	Room temp. (*hν*)	96	[119]
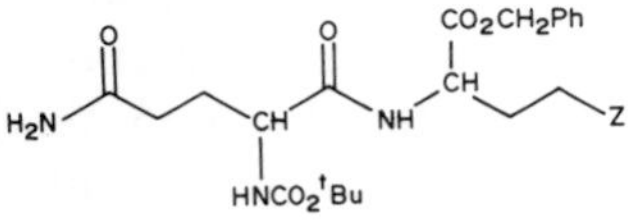	E	Tetrahydrofuran	Room temp. (*hν*)	74	[119]
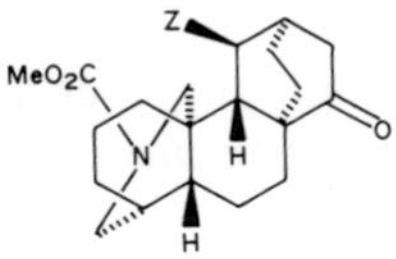	A	Toluene	110	48	[121]
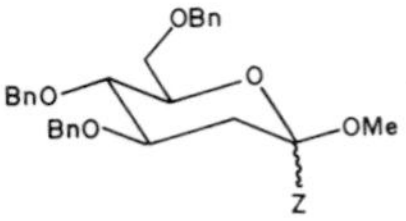	B	Toluene	110	40	[122]

(Z = α,β-CO$_2$H → Z = α-H)

to be preferred since the resultant tertiary alkyl pyridyl disulphide may be removed by simple acid extraction. For aliphatic carboxylic acids, the removal of the carboxyl group can be carried out at room temperature or below using only photolysis from a simple tungsten lamp.

The range of functional groups encountered and tolerated in a variety of complex molecules [123] serves to emphasize the mildness and selectivity of the reaction, and this is illustrated by some selected examples in Table 3.22.

The five general experimental procedures which follow are illustrative not only of the evolution of the sequence in practical terms, but also of the variety of approaches which may be used both in the formation of the acyl thiohydroxamate and in its reductive decarboxylation.

Method A. Reductive decarboxylation using tri-n-butylstannane and in situ *generation of the O-acyl thiohydroxamate from the acid chloride*

The acid chloride (1 mmol) in the appropriate solvent (5 ml) was added to a stirred, dried (Dean–Stark apparatus) suspension of the sodium salt of *N*-hydroxypyridine-2-thione (1.2 mmol) and of 4-dimethylaminopyridine (0.1 mmol) in the same solvent (10 ml) at reflux. After 10–15 min at reflux, tri-n-butylstannane (3 mmol) and AIBN (10–20 mg) in the appropriate solvent (5 ml) were added dropwise over 15 min. The progress of the reaction was monitored by TLC and more stannane and initiator added as necessary. After completion the reaction was cooled to 80°C and treated with tetrachloromethane (10 ml) for 1 h. Subsequently, the reaction mixture was evaporated to dryness and the residue vigorously stirred overnight in a two phase system comprising a saturated solution of iodine in dichloromethane (20 ml) and a saturated aqueous solution of potassium fluoride (20 ml). The white, polymeric precipitate was filtered on celite and washed with dichloromethane. The washings were combined with the reaction mixture and after decantation the aqueous phase was further extracted with dichloromethane (3 × 15 ml). The combined organic phases were washed with sodium thiosulphate (20 ml, 2 M), water (20 ml) and saturated sodium chloride (20 ml), dried on sodium sulphate, filtered and evaporated to dryness to yield the crude reaction product. Pure products were isolated by flash chromatography over silica gel and either recrystallized or distilled (Kugelröhr) as appropriate.

From ref. [118] with permission.

Method B. General procedure for the reductive decarboxylation of carboxylic acids using in situ *generation from the acid chloride and t-butyl mercaptan—normal addition*

The acid chloride (1 mmol) in toluene (5 ml) was added dropwise over a 15 min period to a dried, stirred suspension of the sodium salt of

N-hydroxypyridine-2-thione (1.2 mmol), 4-dimethylaminopyridine (0.1 mmol) and t-butyl mercaptan (0.5 ml, 4.5 mmol) in toluene (20 ml) with heating to reflux using an efficient condenser under a nitrogen atmosphere. The progress of the reaction was monitored by analytical TLC. After completion, the reaction was cooled to room temperature and thoroughly washed with water, then with a saturated solution of sodium chloride. After drying on sodium sulphate, filtration and evaporation to dryness, the products were purified by chromatography on silica.

From ref. [118] with permission.

Method C. Reduction with t-butyl mercaptan—inverse addition

The acid chloride (1 mmol) in toluene (1 ml) was added at room temperature to a stirred solution of N-hydroxypyridine-2-thione (the free thionohydroxamic acid) (1.1 mmol, 140 mg) and pyridine (0.1 ml) in toluene (10 ml) at room temperature. After 10 min, the precipitate was filtered off and the filtrate added dropwise over 30 min to a solution of t-butyl mercaptan (0.5 ml) in toluene (20 ml) at reflux under a nitrogen atmosphere and using an efficient condenser. After completion of the reaction (TLC control) the products were isolated as described in method B.

From ref. [118] with permission.

Method D. Reductive decarboxylation of bridgehead dicarboxylic acid monoesters

A mixture of the carboxylic acid (1 equiv.) and thionyl chloride (2–3 equiv.) was heated under reflux for 2 h. Excess thionyl chloride was removed under vacuum from the cooled mixture and the residue either distilled/sublimed or used further without purification.

A stirred suspension of the sodium salt of N-hydroxypyridine-2-thione (1.1–1.2 equiv.) in anhydrous benzene ($\sim 15\,\mathrm{ml\,g^{-1}}$ of salt) containing t-butyl mercaptan (2.5 equiv.) and 4-dimethylaminopyridine (several milligrams) was deoxygenated and then heated to reflux under an inert atmosphere. This mixture was then treated dropwise with a solution of the acid chloride (1 equiv.) in anhydrous benzene whilst being irradiated with a 300 W lamp. After 2–3 h at reflux, the resultant mixture was cooled to room temperature and treated with an excess of a saturated solution of commercial calcium hypochlorite for 1–2 h ($< 20°C$) before being worked up by method 1, 2 or 3 below.

Method 1. The mixture was filtered at the pump with Celite filter aid and the solid washed with dichloromethane. The aqueous solution was separated and extracted with fresh dichloromethane and the combined organic portions washed with brine and dried ($MgSO_4$). The organic solution was concentrated and the residue treated with a solution of excess potassium hydroxide in 50% aqueous methanol at ambient temperature for 2–3 days. Concentration of the mixture followed by dilution with water and extraction with dichloromethane gave the 2-pyridyl t-butyl disulphide coproduct. The carboxylic acid was isolated by acidification of the aqueous layer and addition of salt followed by extraction with dichloromethane.

Method 2. The mixture was transferred to a separating funnel and washed with successive portions of water until all of the inorganic material was removed. The combined aqueous washings were back-extracted with dichloromethane and the combined organic portions washed with brine, desiccated ($MgSO_4$) and concentrated. The residue obtained was treated as described in method 1 above.

Method 3 (applicable to substrates stable to strong acid).   After removal of the inorganic by-products by either of methods 1 or 2, the 2-pyridyl t-bytyl disulphide coproduct was removed by washing the organic layer with concentrated hydrochloric acid ($\times$ 3), then with water, saturated sodium hydrogen carbonate solution and brine. Desiccation ($MgSO_4$) followed by concentration afforded the product.

From ref. [120] with permission.

Method E.   General procedure for the reductive decarboxylation of amino acid and peptide derivatives

N-Methylmorpholine (1 mmol, 0.11 ml) and isobutyl chloroformate (1 mmol, 0.14 ml) were added at $-15°C$ under nitrogen or argon to a solution of the suitably protected amino acid (1 mmol) in dry tetrahydrofuran (5 ml) in a three necked flask equipped with a thermometer. After 5 min at $-15°C$, a solution of *N*-hydroxy-2-thiopyridone (1.2 mmol, 152 mg) and of triethyl-amine (1.2 mmol, 0.17 ml) in dry tetrahydrofuran (3 ml) was added. The mixture was stirred at $-15°C$ under nitrogen or argon, sheltered from the light (aluminium foil), for about 1 h. (The required ester formation can be followed by TLC (yellow spot, ethyl acetate–hexane (1:1)).) The precipitate of *N*-methylmorpholine hydrochloride was filtered and washed with more dry tetrahydrofuran under aluminium foil protection. The yellow filtrate was irradiated in the presence of *t*-butyl thiol with two 100 W tungsten lamps at room temperature under an inert atmosphere in a water bath until the yellow colour disappeared (usually 10–20 min). The temperature in the flask was close to room temperature. Ether was then added and the organic layer was washed with sodium hydrogen carbonate (0.1 N), water, dilute hydrochloric acid (0.5 N), water again and then with saturated brine. The product was then purified on silica gel.

From ref. [119] with permission.

3.7.3 Decarboxylative rearrangement of *O*-acyl thiohydroxamates [118]

The simplest free radical reaction of *O*-acyl thiohydroxamates is their decarboxylative rearrangement to alkyl-2-pyridyl sulphides [118]. The transformation proceeds via a simple chain mechanism in which reversible addition of the alkyl radical R˙ to the thiocarbonyl group triggers the loss of carbon dioxide and regeneration of a second alkyl radical to propagate the chain. A series of crossover experiments demonstrated that simple tungsten lamp photolysis of the lemon-yellow derivatives from 2-mercaptopyridine-*N*-oxide occurs by a pure chain mechanism, while thermolysis involves, at least in part, a competing cage recombination mechanism [124].

A variety of other thiohydroxamic acids have also been prepared [124,125] which also undergo decarboxylative rearrangement of their corresponding *O*-acyl derivatives (Table 3.23). In contrast to the dihydropyridines, however, more vigorous reaction conditions may be required, as shown.

The derived alkyl-2-pyridyl sulphides are of particular synthetic interest by virtue of their ability to form a chelated lithio anion for reaction with carbon electrophiles (E^+). Reductive removal of the pyridylthio group can then be smoothly accomplished using nickel boride or tri-n-butylstannane.

TABLE 3.23

Reaction conditions for the decarboxylative rearrangement of some *O*-acyl thiohydroxamates

Derivative	Thermal rearrangement (°C)	Photochemical rearrangement (lamp type)
	110	Tungsten
X = Me	110	Mercury, medium pressure
X = Ph	110	Tungsten (slow)
	Stable to 130	Mercury, medium pressure
	Stable to 130	Mercury, medium pressure
	Stable to 130	Mercury, medium pressure

Examination of the examples in Table 3.24 reveals that the rearrangement functions well irrespective of the nature (primary, secondary or tertiary) of the intermediate alkyl radical. Once again, as in the reductive decarboxylation with stannanes or mercaptans, the *O*-acyl thiohydroxamate derivative may be accessed by a variety of routes (Section 3.7), although reaction of the derived acid chlorides with the sodium salt of the thionohydroxamic acid remains the most effective overall procedure [118].

Decarboxylative rearrangement of *O*-acyl thiohydroxamates to alkyl-2-pyridyl sulphides

Substrate (Z = CO₂H)	Solvent	Temperature (°C)	Method	Product (Z = S—py) yield (%)
CH₃(CH₂)₁₄Z	Cyclohexane	81	A	92
	Benzene	20	B	50
	Toluene	110	A	77
	Toluene	110	A	98
	Benzene	80	A	71
Z = β-configuration	Toluene	110	A	72

The following procedures are representative of the thermal and photo-chemical methods.

Method A. General procedure for decarboxylative rearrangement to alkylpyridyl sulphides

The acid chloride (1 mmol) in toluene (5 ml) was added to a stirred, azeotropically dried suspension of N-hydroxypyridin-2-thione sodium salt (1.2 mmol) and 4-dimethylaminopyridine (0.1 mmol) at reflux in toluene. At the end of the reaction, as monitored by TLC, the mixture was cooled to room temperature, filtered on celite, and evaporated to dryness. Chromatography on silica gel gave the pure product.

From ref. [118] with permission.

Method B. Preparation of n-pentadecyl-2-pyridyl sulphide by photolytic rearrangement

Palmitoyl chloride (274 mg, 1 mmol) in benzene (1 ml) was added to a solution of N-hydroxypyridin-2-thione (140 mg, 1.1 mmol) and pyridine (0.01 ml) in benzene (5 ml) with stirring at room temperature. After 20 min at room temperature, the white precipitate of pyridinium hydrochloride was filtered off and the clear yellow filtrate irradiated at room temperature with a 300 W tungsten lamp for 45 min. Evaporation of the solvent followed by chromatography on silica gel gave first n-pentadecane (49 mg, 23%) and then n-pentadecyl-2-pyridyl sulphide (159 mg, 50%).

From ref. [118] with permission.

3.7.4 Decarboxylative halogenation ($RCO_2H \rightarrow RY$, Y = Cl, Br, I) [118, 126]

Trapping of the alkyl radical derived by fragmentation of an *O*-acyl
thiohydroxamate by a halogen atom donor has led to an excellent alternative
to the classical Hunsdiecker reaction and its variants which totally avoids
the need for heavy metal salts and powerful electrophilic reagents. The trichloromethyl radical ($\cdot CCl_3$) is the chain carrier $X^\cdot$ (Section 3.7.1, p. 86)
of choice in these reactions for the production of nor-chlorides and bromides,
requiring only the appropriate selection of tetrachloromethane or bromotrichloromethane as the solvent, and initiation of the reaction either by
tungsten lamp photolysis or by heating to reflux. Analogous decarboxylative
iodination is best achieved using iodoform as the reagent either in benzene,
or, preferably, cyclohexene. In this way excellent yields of nor-halides have
been obtained from a variety of variously functionalized, primary, secondary
and tertiary alkyl substituted carboxylic acid derivatives (Table 3.25).
Particularly challenging applications of the decarboxylative halogenation
sequence include such substrates as the sensitive triene acid shown, where
all classical electrophilic Hünsdiecker variants were unsuccessful [127], and
the preparation of the fragile α-halooxetane by the Fleet group [128], as well
as other successful reports [129].

$Z = CO_2H \rightarrow Z = Br (75\%)$ $Z = CO_2^- Na^+ \rightarrow Z = Cl (18\%)$

In the case of vinyl and aromatic carboxylic acids, the usual problems of
induced homolysis are encountered; the overall efficiency of the chain
sequence is reduced and more vigorous conditions were originally required.
In preparative terms, however, these problems are simply overcome by further
addition of significant quantities of the initiator [130,131]. The advantages
of such a method include the ability to operate on both electron rich and
electron poor aromatic systems, and to access synthetically useful aryl iodides
for metal halogen exchange.

Accordingly, the required experimental conditions for a given transformation may be conveniently grouped into four sections as a function of the
nature of the substrate acid and the halogen atom required.

The appropriate general method may then be selected from those outlined
below.

General method A. Decarboxylative chlorination (or bromination)
of aliphatic and alicyclic carboxylic acids

The acid chloride (1 mmol) in tetrachloromethane (or bromotrichloromethane) (5 ml) was added dropwise over a 15 min period to a stirred, dried

TABLE 3.25

Decarboxylative halogenation of carboxylic acids via O-acyl thiohydroxamates

Substrate (Z = COCl)	Solvent and/or halogenating agent	Temperature (°C)	General procedure	Product (Z = halogen) yield (%)	Ref.
$CH_3(CH_2)_{14}Z$	CCl_4	76	A	70	[118]
$CH_3(CH_2)_{14}Z$	$BrCCl_3$	105	A	95	[118]
$CH_3(CH_2)_{14}Z$	Benzene + CHI_3	80	B	74	[118]
$CH_3(CH_2)_{14}Z$	Cyclohexene + CHI_3	83	B	97	[118]
$(PhCH_2)_2CHZ$	CCl_4	76	A	72	[118]
$(PhCH_2)_2CHZ$	$BrCCl_3$	105	A	90	[118]
$(PhCH_2)_2CHZ$	Cyclohexene + CHI_3	83	B	66	[118]
adamantyl–Z structure	$BrCCl_3$	105	A	98	[118]
steroid (AcO-substituted) –Z structure	$BrCCl_3$	105	A	71 (on a 9 g scale)	[132]
BocNH–CH(CO₂Bn)–CH₂CH₂–Z structure	$BrCCl_3$	105	A	82	[119]
2,4,6-trimethoxyphenyl–Z structure	$BrCCl_3$	100	C	62	[131]
4-OBn-3-OMe-phenyl–Z structure	$BrCCl_3$	100	C	55	[131]
2-Cl-3-NO₂-phenyl–Z structure	$BrCCl_3$	100	C	43	[131]
1-naphthyl–Z structure	Toluene + CHI_3	100	D	69	[131]

TABLE 3.25 (*continued*)

Substrate (Z = COCl)	Solvent and/or halogenating agent	Temperature (°C)	General procedure	Product (Z = halogen) yield (%)	Ref.
O$_2$N-⬡-Z	Toluene + CHI$_3$	110	D	54	[131]
MeO, MeO, OMe -⬡-Z	Toluene + CHI$_3$	110	D	31	[131]
	Toluene + CH$_2$I$_2$	110	D	40	[131]

suspension of the sodium salt of *N*-hydroxypyridin-2-thione (180 mg, 1.2 mmol) and 4-dimethylaminopyridine (12 mg, 0.1 mmol) in tetrachloromethane (or bromotrichloromethane) (10 ml) at reflux under a nitrogen atmosphere. The reaction was monitored by TLC and after completion was cooled to room temperature, filtered on celite and evaporated to dryness. The crude product thus obtained was purified by chromatography on silica gel to yield the norchloride (or bromide) and 2-pyridyltrichloromethyl sulphide.

From ref. [118] with permission.

General method B. Decarboxylative iodination of aliphatic and alicyclic carboxylic acids

The acid chloride (1 mmol) in the appropriate solvent (see Table 3.25) (1 ml) was added to a dried stirred suspension of the sodium salt of *N*-hydroxypyridin-2-thione (165 mg, 1.1 mmol), 4-dimethylaminopyridine (12 mg, 0.1 mmol) and iodoform (433 mg, 1.1 mmol) at reflux in the approximate solvent under a nitrogen atmosphere. At the end of the reaction (TLC control), the reaction mixture was cooled to room temperature, filtered on celite and concentrated to dryness. The pure products were isolated by chromatography on silica gel.

From ref. [118] with permission.

General method C. Decarboxylative bromination of aromatic carboxylic acids

To a suspension of the sodium salt of *N*-hydroxypyridin-2-thione (1.1 mmol) in refluxing bromotrichloromethane (5 ml) under an inert atmosphere was added dropwise, over a 30 min period, a solution of the aromatic acid chloride (1 mmol) and AIBN (~ 25 mg) in the same solvent (5 ml). After a further heating period of 5 min, the cooled reaction mixture was evaporated under reduced pressure and the residue purified by chromatography on silica gel using dichloromethane–pentane mixtures.

From ref. [131] with permission.

General method D. Decarboxylative iodination of aromatic carboxylic acids

To a suspension of the sodium salt of *N*-hydroxypyridin-2-thione (1.1 mmol) in refluxing dry toluene (3 ml) was added, under an inert atmosphere, iodoform (or diiodomethane) (1.1 mmol) followed by a solution of the acid chloride (1 mol) and AIBN (~ 25 mg) in the same solvent (5 ml) over a 30 min period. After further heating for 5 min, the reaction mixture was cooled to room temperature and the solvent evaporated under reduced pressure. The residue was purified by chromatography on silica gel using dichloromethane–pentane mixtures.

From ref. [131] with permission.

3.7.5 Decarboxylative chalcogenation ($RCO_2H \rightarrow RSR'$, $RSeR'$, $RTeR'$)

The reaction of disulphides, diselenides or ditellurides with *O*-acyl thiohydroxamates provides an efficient method for the synthesis of unsymmetrical thio-, seleno- and telluroethers, respectively. Initial experiments [133] under thermal conditions in refluxing toluene, or even in molten disulphide, revealed that the decarboxylative rearrangement to alkyl pyridyl sulphides (Section 3.7.3) was an unfavourable competitive process which could only be effectively eliminated by using a vast excess of the dichalcogenide to trap the intermediate alkyl radical R˙ in the S_H2 process. Fortunately, further studies demonstrated that low temperature photolysis using a simple tungsten lamp led to clean conversions, even using only a slight excess of reagents [134].

In addition to ready access to a wide variety of primary, secondary and tertiary phenyl sulphides and tellurides, the Barton group also developed

TABLE 3.26

Photochemical decarboxylative chalcogenation of carboxylic acids via O-acyl thiohydroxamates

Substrate (Z = CO$_2$H)	Dichalcogenide	Product	Yield (%)	Ref.
CH$_3$(CH$_2$)$_{14}$Z	PhSSPh	Z = SPh	82	[134]
	PhSeSePh	Z = SePh	85	[134]
	(naphthyl-Te)$_2$	Z = naphthyl-Te	82	[134]
steroid (AcO)	(PhO-C$_6$H$_4$-Te)$_2$	Z = Te-p-C$_6$H$_4$OPh	55	[134]
adamantane	PhSeSePh	Z = SePh	93	[134]
	(MeO-C$_6$H$_4$-Te)$_2$	Z = Te-C$_6$H$_4$-OMe	94	[134]
triterpene (AcO, OAc)	MeSeSeMe	Z = SeMe	71	[134]
steroid (AcO)	NCSeSeSeCN	Z = SeCN	77	[134]
PhCH$_2$O...NH...CO$_2$CH$_2$Ph	PhCH$_2$SeSeCH$_2$Ph	Z = SeCH$_2$Ph	64	[135]
	NCSeSeSeCN	Z = SeCN	73	[135]

concise syntheses of two of the most important selenoamino acids, L-selenomethionine and L-selenocystine, starting from commercially available glutamic and aspartic acid derivatives [135]. These and other representative examples are shown in Table 3.26.

In addition to the use of simple dialkyl and diaryl diselenides, the use of crystalline dicyanogen triselenide for introduction of the selenocyanate moiety is also noteworthy [135].

A detailed general procedure *via* the acid chloride method, and a specific example of the chloroformate method for the preparation of *O*-acyl thiohydroxamates, are given below.

General procedure for decarboxylative chalcogenation

Preparation of the acid chlorides. The acid chlorides used in the following preparations of the mixed anhydrides were prepared immediately before use and were not purified. Thus, to a solution of the corresponding carboxylic acid (1 mmol) in benzene (5 ml) was added oxalyl chloride (3 mmol) and a drop of dimethyl formamide. After stirring for 2 h at room temperature with protection from moisture, the solvent and excess oxalyl chloride were evaporated and the residual acid chloride used as such.

Synthesis of O-*acyl thiohydroxamates.* (*Note:* These compounds are sensitive to light. The reaction vessel, chromatography column, etc., should therefore be covered with aluminium foil.) To a solution of the acid chloride (10 mmol) in dry, degassed dichloromethane (50 ml) was added the sodium salt of *N*-hydroxypyridin-2-thione (10.5 mmol). After stirring at room temperature in the absence of light and under an inert atmosphere for 1–2 h, the reaction mixture was rapidly filtered and the solvent evaporated without heating. The yellow residue may be used as such or further purified by filtration on a short silica gel column.

Synthesis of sulphides, selenides and tellurides. The *O*-acyl thiohydroxamate (1 mmol) was added to an ice cold, degassed solution of the dichalcogenide (2 mmol) in dichloromethane (20 ml) under argon. Irradiation for 20 min with a 300 W tungsten lamp (projector lamp) placed near the reaction vessel followed by concentration and chromatography of the residue on silica afforded the desired product. In some cases involving diphenyl diselenide, treatment with aqueous sodium borohydride removed the excess diphenyl diselenide and simplified the purification.

From ref. [134] with permission.

Preparation of benzyl N-t-butoxycarbonylselenomethionate

To a stirred solution of the protected glutamic acid derivative (337 mg, 1 mmol) in dry tetrahydrofuran were added with cooling to $-15°C$, and under nitrogen, N-methylmorpholine (0.11 ml) and isobutyl chloroformate (0.14 ml). After 15 min at this temperature, a cooled solution of N-hydroxypyridin-2-thione (152 mg) and triethylamine (0.17 ml) in dry tetrahydrofuran was added. The mixture was stirred in the dark for 20 min and then filtered rapidly. To the filtrate containing the O-acyl thiohydroxamate was added dimethyl diselenide (1.86 g) and the mixture was then irradiated with a 300 W tungsten lamp under nitrogen for 20 min. Subsequently, the reaction mixture was diluted with ether, and the organic phase washed successively with water, then brine, and dried over sodium sulphate. Removal of the solvent and chromatography on silica gel gave the title compound as an oil (300 mg, 78%), $[\alpha]_D^{20} - 34°$ ($c = 1.1$ in methanol).

From ref. [135] with permission.

3.7.6 Decarboxylative phosphonylation $(RCO_2H \rightarrow RPO(SPh)_2)$ [136]

The replacement of a carboxylic acid group in a biologically active molecule by a phosphonic acid moiety is a useful transformation, which can be realized simply by reaction of an O-acyl thiohydroxamate with tris(phenylthio)-phosphorus.

The reaction sequence begins with the simple free radical chain sequence involving the phenythiyl radical as the chain carrier X (Section 3.7.1) and trapping of the generated alkyl radical R· at the phosphorus centre.

The initial trivalent alkylbis(phenylthio)phosphine, however, then reacts with the unsymmetrical disulphide also formed in the free radical chain reaction to give an equilibrium mixture containing the pentavalent intermediate, which on hydrolytic work up is driven to completion to give the isolated *S,S*-diphenyl dithiophosphonate. Although yields are, in general, moderate, the overall transformation is a useful one, especially for primary carboxylic acids. The mixed anhydrides derived from *N*-hydroxypyridin-2-thione can also be employed. Several examples are collected in Table 3.27.

3.7.7 Decarboxylative sulphonation ($RCO_2H \rightarrow RSO_2Spy$) [137]

The ability of sulphur dioxide, when used in sufficiently large excess, to act as an efficient and kinetically favourable trap for alkyl radicals (R·), allows the insertion of an additional step into the propagative chain sequence, without interference from the decarboxylative rearrangement (Section 3.7.3).

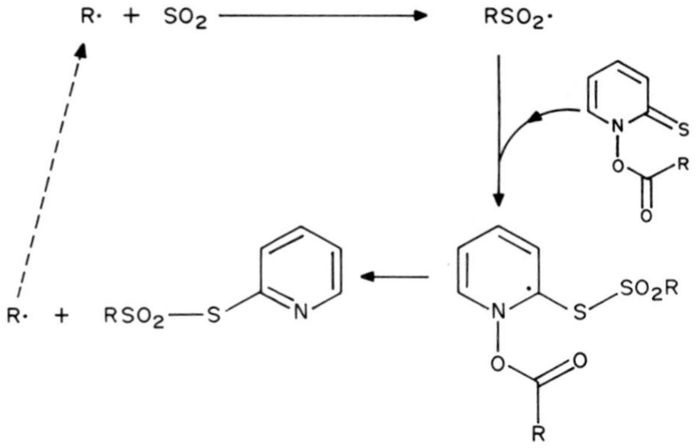

R· + SO$_2$ ⟶ RSO$_2$·

R· + RSO$_2$—S

TABLE 3.27

Preparation of S,S-diphenyl dithiophosphonates from carboxylic acids via O-acyl thiohydroxamates [136]

Substrate (Z = CO$_2$H)	Thiohydroxamic acid[a]	Product yield (%) (Z = PO(SPh)$_2$)
C$_{15}$H$_{31}$Z	A	67
Ph$_2$CHCH$_2$Z	A	64
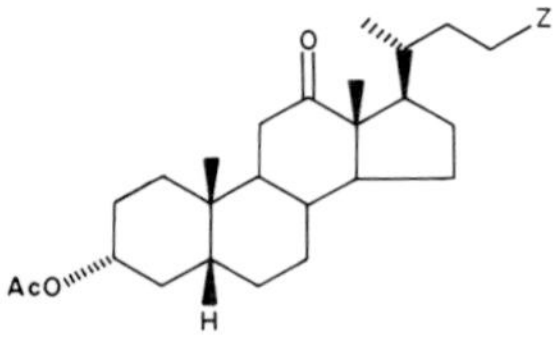	A	60
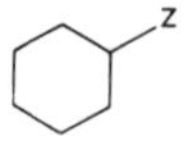	A	50
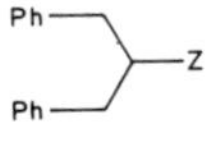	B	35
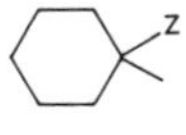	B	26

[a] Reactions were run at room temperature using thionohydroxamic acids

A and B in chlorobenzene and dichloromethane,

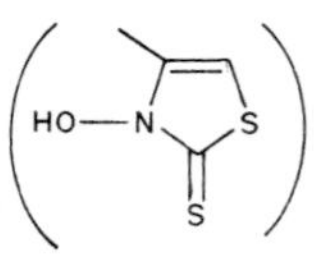

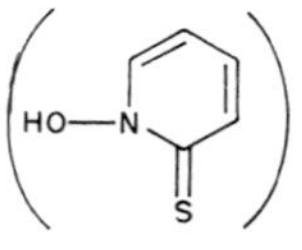

respectively, under a nitrogen atmosphere containing a small trace of oxygen as the initiator.

The resulting S-pyridyl alkylthiosulphonates are useful precursors for a variety of ionic transformations including a non-oxidative preparation of unsymmetrical sulphones and sulphonamides. For this latter transformation, in addition to a conventional route *via* the sulphonyl halide, the Barton group also developed an ingenious method for more sensitive substrates which involved stirring together in dimethylformamide a mixture of the thiosulphonate, the amine and 1,2-dibromotetrachloroethane. The following experimental procedure is representative. (*O*-Acyl thiohydroxamates are prepared either from the derived acid chloride or *via* the chloroformate method described in Section 3.7.5.)

General method for the preparation of thiosulphonates by decarboxylative sulphonylation of O-acyl thiohydroxamates

Dry, degassed dichloromethane (30 ml) was cooled to $-40°C$ in a Schlenk tube under an inert atmosphere. Sulphur dioxide was then condensed into the tube until a total volume of about 40 ml was obtained. After warming to $-10°C$, the O-acyl thiohydroxamate (5 mmol) was added and the mixture stirred for a few minutes at this temperature. The cooling bath was removed and the resulting solution was irradiated using a 250 W tungsten projector lamp cooled with a stream of air until the yellow colour of the mixed anhydride was discharged (~ 30 min). The solvent and excess sulphur dioxide were then evaporated and the residue purified by chromatography on a short column of silica gel using dichloromethane as the eluant, followed by crystallization.

From ref. [137] with permission.

The yields for several representative examples are shown in Table 3.28.

TABLE 3.28

Decarboxylative sulphonylation of O-acyl thiohydroxamates [137]

Substrate ($Z = CO_2H$)	Product $\left(Z = SO_2S{-}\underset{N}{\bigcirc}\right)$ yield (%)
$CH_3(CH_2)_{13}CH_2Z$	91
$PhCH_2CH_2Z$	65
	90
	85
	38
	30
	54

3.7.8 Decarboxylative hydroxylation (RCO$_2$H → ROH)

The conversion of a carboxylic acid to its derived nor-alcohol generally requires several steps by conventional methods. The chemistry of O-acyl thiohydroxamates, however, has led to the invention of two mechanistically different radical variants which achieve this transformation in useful yields.

In the first of these [118, 138], as in decarboxylative sulphonylation (Section 3.7.7), kinetic considerations permit the insertion of an additional step into the propagative chain sequence, *viz.* interception of the carbon centred radical by oxygen is some 10 000 times faster than hydrogen atom capture from a mercaptan.

The intermediate hydroperoxy radical thus formed can then undergo hydrogen atom capture from the mercaptan in order to release the initial product hydroperoxide and the chain carrying thiyl radical. The use of a non-nucleophilic thiol was found to be essential for maintenance of clean reactions and the obtention of high yields.

Although the hydroperoxide may, if desired, be isolated in moderate to good yields, further reduction to the nor-alcohol can be carried out using trimethyl phosphite or dimethyl sulphide. Alternatively, for hydroperoxides derived from primary or secondary carboxylic acids, reaction with tosyl chloride and pyridine may be used to give the derived aldehyde or ketone respectively.

A selection of aliphatic and alicyclic primary, secondary and tertiary carboxylic acids have been converted to their derived nor-alcohols in acceptable overall yields by this straightforward and readily applicable procedure (Table 3.29).

Initially, experiments were carried out in oxygen saturated toluene solutions at 80°C using either *in situ* O-acyl thiohydroxamate formation (method A) or the preformed ester (method B). Room temperature photolysis using a tungsten lamp (method C) is also very convenient. The overall procedures for nor-alcohol formation are set out below.

General methods for the formation of nor-hydroperoxides

Method A. The acid chloride (1 mmol) and t-butyl mercaptan (generally 9.0 mmol), both in toluene (10 ml) were added simultaneously to a stirred

TABLE 3.29

Decarboxylative hydroxylation of carboxylic acids via O-acyl thiohydroxamates using t-butyl mercaptan and oxygen followed by reductive work up [118]

Substrate (Z = CO_2H)	Method	Work up procedure	Nor-alcohol (Z = OH) yield (%)
$CH_3(CH_2)_{14}Z$	A	$(MeO)_3P$	75
	A	Me_2S	51
	C	Me_2S	57
	C	$(MeO)_3P$	67
	C	—	77[a] (Z = OOH)
(steroid structure, AcO substituents, Z side chain)	A	$(MeO)_3P$	69
(triterpene structure, AcO, OAc substituents, Z)	A	—	56 (Z = 17β-OH), 33 (Z = 17β-OOH)
(Ph—, Ph— structure with Z)	B	$(MeO)_3P$	82

[a] Yield estimated by NMR.

dried suspension of the sodium salt of *N*-hydroxypyridin-2-thione (180 mg, 1.2 mmol) and 4-dimethylaminopyridine (12 mg, 0.1 mmol) in toluene (10 ml) at 80°C and through which oxygen was being passed via a sinter at a rate of approximately 0.33 l min^{-1}. Decoloration and TLC analysis of the normally yellow solution indicated complete reaction.

Method B.   The acid chloride (1 mmol) in toluene (2 ml) was added at room temperature to a stirred solution of *N*-hydroxypyridin-2-thione (140 mg, 1.1 mmol) and pyridine (0.25 ml) in toluene (10 ml) at room temperature. After 10–15 min, the precipitated pyridinium hydrochloride was removed by filtration and the yellow filtrate added dropwise to a solution of t-butyl mercaptan (9.0 mmol) in toluene that was being continually saturated with oxygen as in method A. The *O*-acyl thiohydroxamate can also be preformed by the chloroformate method (Section 3.7.5).

Method C.   A yellow solution of the ester (1 mmol) in toluene (10 ml) was prepared as in method B and added dropwise with irradiation by a 300 W tungsten lamp over 15 min to a stirred solution of t-butyl mercaptan (9.0 mmol) in toluene (10 ml) that was being continually saturated with oxygen, as in method A, at room temperature.

General method for the reduction of nor-hydroperoxides to nor-alcohols

After formation of the hydroperoxides by either of the methods A, B or C outlined above, the crude reaction mixture was reduced either with dimethyl sulphide (typically 1–2 ml per millimole of carboxylic acid substrate) at 80°C for 1 h or with trimethylphosphite (0.25 ml) at room temperature for 2.0 h. After complete reduction, the reaction mixture was thoroughly washed with water (3 × 50 ml), dried on sodium sulphate, filtered and evaporated to dryness. Purification by chromatography on silica gel effected separation of the nor-alcohol from t-butyl-2-pyridyl disulphide.

From ref. [118] with permission.

For the fragmentation of hydroperoxy tosylates to their corresponding carbonyl compounds, method C has proven to be the most effective, since all previously used reagents were eliminated prior to photolysis. Simple removal of the excess mercaptan by water washing is all that is required before treatment with tosyl chloride and pyridine. The yields for a variety of hydroperoxides, as estimated by NMR, together with several representative conversions to the corresponding oxo derivatives, are shown in Table 3.30, and the following general procedure may be employed.

TABLE 3.30

Formation of nor-hydroperoxides and nor-oxo derivatives from carboxylic acids via
O-acyl thiohydroxamates [118]

Substrate (Z = CO_2H)	Hydroperoxide (Z = OOH) yield (%)[a]	Carbonyl compound	Yield (%)
(cyclooctenyl)-Z	45[b,c]	—	
$CH_3(CH_2)_{14}Z$	89	$CH_3(CH_2)_{13}CHO$	53
(steroid, AcO...)-Z	74	(steroid, AcO...)-CHO	56
Ph—Ph—Z	85	Ph—Ph—=O	62

[a]Estimated by NMR. [b]Isolated yield. [c]Chapter 7, ref. [124].

General method for the fragmentation of hydroperoxides with tosyl chloride and pyridine

The hydroperoxide was prepared by method C above. After aqueous work up the ^1H NMR of the crude reaction mixture showed it to be a 1:1 mixture of t-butyl-2-pyridyl disulphide and hydroperoxide, thereby allowing an estimation of hydroperoxide yield. The mixture was dissolved in pyridine (2 ml) and treated with tosyl chloride (250 mg, 1.3 mmol) under magnetic stirring. After 3 h at room temperature, the reaction mixture was diluted with 2 M hydrochloric acid (20 ml) and extracted with dichloromethane (20 ml). The organic phase was successively washed with water (20 ml) then brine (20 ml), dried over sodium sulphate, filtered, and the filtrate evaporated to dryness. Further purification was carried out by using chromatography on silica gel.

From ref. [118] with permission.

A second method [139] for the conversion of carboxylic acids into nor-alcohols was discovered by the Barton group in exploring the reaction

TABLE 3.31

Preparation of nor-alcohols from *O*-acyl thiohydroxamates and tris(phenylthio)-
antimony in the presence of oxygen and water [139]

Substrate (Z = CO₂H)	Thiohydroxamate[a]	Solvent	Time (h)	Nor-alcohol (Z = OH) yield (%)
CH₃(CH₂)₁₄Z	A	Et₂O	4	85
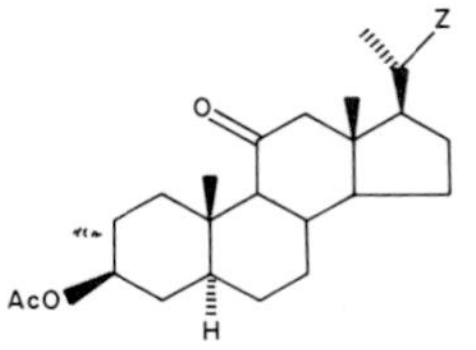	A	PhCl	12	85
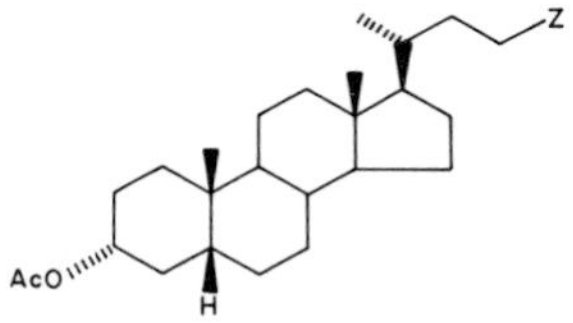	A	PhCl	12	90
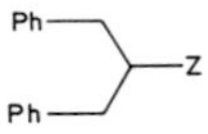	A	PhCl / Et₂O	12 / 12	91 / 85
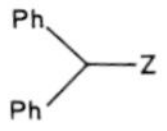	A	Et₂O	12	93
	B	Et₂O	0.5	91
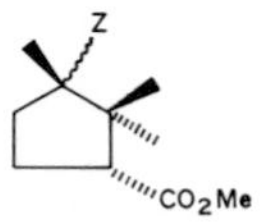	A	Et₂O	2	82
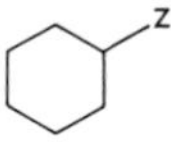	A	CH₂Cl₂	12	91

[a] A = structure shown; B = structure shown.

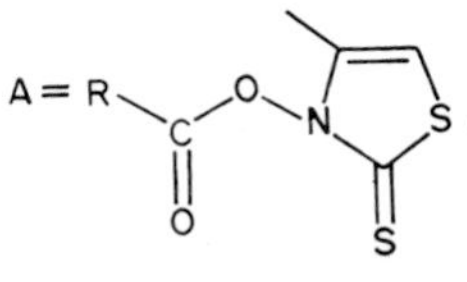

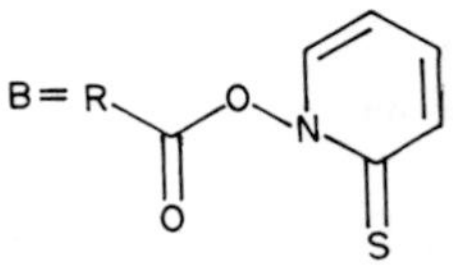

of *O*-acyl thiohydroxamates with tris(phenylthio)antimony. As in the case of decarboxylative phosphonylation with the congeneric phosphorus derivative (Section 3.7.6), the initial sequence is the standard decarboxylative chain reaction in which the phenylthiyl radical acts as an efficient carrier.

The initially formed organoantimony intermediate, however, undergoes ready autoxidation, and the resultant peroxide, after rearrangement and reductive elimination of diphenyl disulphide gives an alkoxide from which the parent nor-alcohol may be liberated by simple hydrolysis.

The experimental procedure [139] given consists of simply stirring the *O*-acyl thiohydroxamate derivative with tris(phenylthio)antimony at room temperature in an open flask under air, and thereby allowing oxygen and adventitious moisture to perform the task. Excellent yields of nor-alcohols have thus been obtained from a variety of primary, secondary and tertiary carboxylic acids in a reaction which is probably easier to carry out and better adapted to large scale work than the procedure via the hydroperoxide.

Several representative examples are shown in Table 3.31.

3.8 FUNCTIONAL GROUP INTERCHANGE USING ORGANOMETALLIC DERIVATIVES OF BORON, MERCURY AND COBALT

3.8.1 Introduction

The induced homolysis of readily available and highly functionalized organoboron [140] and organomercury [141] compounds provides an extremely mild and useful method for the generation of carbon centred radicals. Given that the chain reactions of the boron derivatives with electron deficient olefins were described by H. C. Brown some 20 years ago [142], this is a remarkable example of an "induction period" prior to their recent renaissance and subsequent appreciation by the chemical community.

More recently, the stimulus provided by interest in the mode of action of coenzyme vitamin B_{12} has led to the acknowledgement of organocobalt reagents [143] as useful precursors for homolysis. In particular, model corrinoids in which nature's tetraazamacrocycle has been replaced by chelating equatorial ligands such as bis(dimethylglyoxime) (dmgH) and N,N-ethylenebis(salicylideneimato) (salen) have been most often used because of their availability by nucleophilic substitution of an organic halide or tosylate with the corresponding cobalt(I) complexes [144].

RCo(III)(dmgH)$_2$(py)   RCo(III)salen(py)

Typical methods for the generation of alkyl radicals (R.) are set out here.

Since the major use of the above metal and metalloid derivatives has been in carbon–carbon bond forming reactions (Chapters 6 and 7) their discussion in the present section is relatively brief and restricted to some of the more recent bimolecular homolytic substitution reactions which lead to functional group interchange.

3.8.2 Preparation of alcohol derivatives ($R_xML_n \rightarrow ROH$)

3.8.2.1 From organoboranes and organomercury compounds

In view of the standard oxidative work up of organoboranes using basic hydrogen peroxide, it is often forgotten that the autoxidation of trialkylboranes is also a very useful preparative method in its own right and involves a highly efficient chain reaction [145]. Aspects of the selectivity in this reaction have recently been discussed [146].

In a similar fashion, the free radical chain reaction of molecular oxygen with alkyl mercuric hydrides [147], generated by *in situ* borohydride reduction of the corresponding halides, is a proven demercuration reaction which continues to find routine application in organic synthesis.

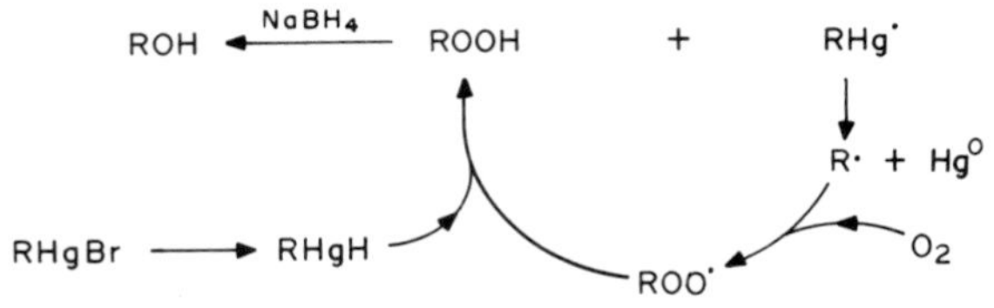

ROH ←—NaBH₄— ROOH + RHg·

RHgBr —→ RHgH

R· + Hg⁰

ROO·   O₂

As anticipated, the rate of capture of the 5-hexenyl radical by molecular oxygen is significantly faster than the rate of 5-*exo* trigonal ring closure [148].

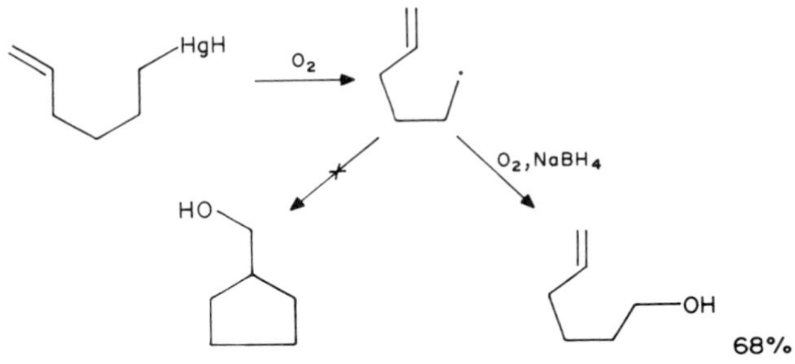

HgH   O₂

O₂, NaBH₄

HO

OH

68%

Some recent examples of hydroxylative demercuration include the applications shown in the prostaglandin [149] and cyclic aminoalditol [150] areas.

The preparative simplicity of the overall sequence is well illustrated by the preparation of the oxazine precursors used in a stereoselective synthesis of γ-hydroxy-α-amino acids [151]. In this case, intramolecular aminomercuration is followed by hydroxylative demercuration.

cis-, trans-3-Benzyloxycarbonyl-4-hydroxymethyl-6-methyl-tetrahydro-1,3-oxazine

The alkene (0.423 g, 1.7 mmol) was dissolved in 10 ml of acetonitrile with sodium hydrogen carbonate (146 mg, 1.7 mmol). Mercuric nitrate (827 mg, 2.5 mmol) was added to the reaction solution, and the mixture was stirred for 1 h at room temperature. Concentrated aqueous potassium bromide (5 ml) was added and the mixture was stirred vigorously for 2 h. Ethyl acetate (20 ml) was added to the mixture to dilute the solution. The organic phase was separated from the aqueous phase, and the aqueous phase was extracted with 10 ml of ethyl acetate. The combined organic layer was concentrated under reduced pressure to give 0.897 g (> 98%) of organomercurial bromide. Analysis by ^{1}H NMR showed a 3:1 ratio. Dimethylformamide (3.0 ml) and sodium borohydride (13 mg) were placed in a 25 ml centrifuge tube, which was capped with a rubber septum containing an inlet and an outlet needle for oxygen. Oxygen was supplied to the bottom of the centrifuge tube. The oxygen flow was controlled at a rate of 300–400 ml min^{-1}. The sodium borohydride mixture was flushed with oxygen for 20 min. The oxidative demercuration was accomplished by adding 10 ml of dimethylformamide containing the organomercurial bromide (0.130 g, 0.23 mmol) slowly over a 15 min period to the sodium borohydride solution via a syringe pump. Elemental mercury precipitated immediately. Aqueous sulphuric acid (0.1 N, 10 ml) was added to the reaction mixture. The mixture was stirred for 1 h at room temperature, and the organic phase was extracted twice with 50 ml of ether. The combined organic layer was dried over anhydrous magnesium sulphate and concentrated under reduced pressure to give 58 mg of product mixture as an oil, which was purified by preparative TLC (40% ethyl acetate–hexane). The *cis*-alcohol was collected in 60% yield (36 mg) and the *trans*-alcohol was collected in 20% yield (12 mg).

From ref. [151] with permission.

3.8.2.2 From organocobalt derivatives

The "insertion" of molecular oxygen into the cobalt–carbon bond is a highly favourable process which can be simply carried out under photochemical conditions using the light from a standard 100 W tungsten lamp. Reduction of the resultant peroxycobalt complex may then by achieved with sodium borohydride [152, 153].

Recent examples of this reaction include the preparation of a wide range of hydroxyl substituted aromatic and heterocyclic molecules by the Pattenden group [153] as one possible component of their more general cobalt initiated radical cyclization trap–functional group interconversion sequence (Chapter 7).

The same group emphasizes, however, that the process involving non-chain carbon radical trapping with 2,2,6,6-tetramethylpiperidine-N-oxyl (TEMPO) is a more practical procedure and this would certainly seem to be reflected in the isolated yields for the case of the above dihydrobenzofuran [154].

An elegant catalytic cycle based on hydridocobaltation of styrene derivatives and using meso(tetraphenylporphyrinato)cobalt(II) has also been developed [155].

TABLE 3.32

Oxidation of styrene derivatives with oxygen and tetraethylammonium borohydride catalysed, by meso-(tetraphenylporphyrinato)cobalt(II) [155]

Alkene	Product	Yield (%)
Ph⌄⌄	OH, Ph⌄	87
Cl⌄(aryl ring)⌄	Cl⌄(aryl ring)⌄OH	92
NO₂(aryl ring)⌄	NO₂(aryl ring)⌄OH	79
(phenyl ring)	(phenyl ring)OH	94
⌄⌄⌄⌄(phenyl ring)	⌄⌄⌄OH(phenyl ring)	68

The yields for a variety of usefully functionalized styrene derivatives are shown in Table 3.32 and were obtained following the procedure outlined below.

General method for the catalytic hydration of styrene derivatives

A mixture of olefin (1 mmol), meso(tetraphenylporphyrinato)cobalt(II) (Co(TPP)) (9×10^{-3} mmol) and Et_4NBH_4 (0.5 mmol) was stirred under air in 5 ml of a 50% mixture of 1,2-dimethoxyethane and 2-propanol at room temperature. Stirring was continued with an occasional addition of BH_4^- until all the substrate was consumed. When the reaction was completed, 15 ml of water was added to the reaction mixture and the precipitated catalyst was removed by filtration. The filtrate was extracted by ether, and the organic layer was dried over Na_2SO_4. The product alcohol was obtained in an almost pure form after evaporation of the solvent.

From ref. [155] with permission.

3.8.3   Chalcogenation (RML$_n$→RSR', RSeR', RTeR')

3.8.3.1 From organomercury compounds

The reaction of organomercury halides (RHgCl) with a variety of suitable dichalcogens proceeds readily either under photochemical conditions or using AIBN as a thermal initiator to give the derived alkyl aryl sulphides, selenides or tellurides [156–158].

$$R\!-\!HgCl \xrightarrow[\text{hv or AIBN}]{X-Y} R\!-\!X$$

$$X\!-\!Y$$

PhS—SPh
PhSe—SePh
PhSe—SO$_2$Ph
PhSe—SCOPh
PhSe—CN
PhTe—TePh

A variety of primary and secondary alkyl derivatives have been studied, and functionalized organomercurials, available by solvomercuration, also undergo efficient reaction [159]. The light induced reaction of vinyl mercuric halides with diselenides or ditellurides, which may be considered to proceed via an addition–elimination reaction, also provides a useful route to (1-alkenyl)phenyl selenides and tellurides [157, 160, 161]. A variety of representative examples are shown in Table 3.33.

A particularly noteworthy example of this sequence has been described by McMurry [162, 163] for a highly functionalized decalin derivative; in this case the formation of the new carbon selenium bond proceeds without ring opening of the intermediate cyclopropyl carbinyl radical ($k = 10^8\,\text{s}^{-1}$). Selenoxide elimination then provides the olefin.

Typical experimental procedures for the preparation of alkyl phenyl chalcogenides are given below.

TABLE 3.33

Preparation of alkyl phenyl sulphides, selenides and tellurides by photoinitiated reaction of alkylmercuric halides in the presence of a chalcogen donor

Substrate (Z = HgX, X = halogen)	X	Reagent	Product	Yield (%)	Ref.
$CH_3(CH_2)_4CH_2Z$	Cl	PhSeSePh	Z = SePh	82	[156]
	Cl	PhSeSO2—(aryl)	Z = SePh	82	[156]
	Cl	PhTeTePh	Z = TePh	83	[156]
$(CH_3)_3CCH_2Z$	Cl	PhSeSePh	Z = SePh	86	[156]
	Cl	PhSeSO2—(aryl)	Z = SePh	75	[156]
	Cl	PhTeTePh	Z = TePh	78	[156]
	Cl	PhSSPh	Z = SPh	74	[156]
(bicyclic structure, Z)	Br	PhSeSePh	Z = SePh	53	[156]
	Br	PhSeSO2—(aryl)	Z = SePh	48	[156]
	Br	PhTeTePh	Z = TePh	45	[156]
(cyclohexane, OCH3, Z)	Br	PhSeSePh	Z = SePh	95	[159]
$CH_3(CH_2)_6CH_2CH(OH)CH_2Z$	Br	PhSeSePh	Z = SePh	95	[159]
	Br	PhSeSCOPh	Z = SePh	92	[159]
	Br	PhSeCN	Z = SePh	80	[159]
(Ph/H/H/Z alkene)	Cl	PhSeSePh	Z = SePh	90	[160]
(alkene, H/H/Z)	Cl	PhSSPh	Z = SPh	100	[164]

General method for the preparation of alkyl phenyl selenides from alkylmercuric halides and their oxidative conversion to olefins

In a representative experiment, organomercuric chloride (17 mg, 0.034 mmol) and diphenyl diselenide (21 mg, 0.068 mmol) were slurried in benzene (0.5 ml) and the yellow mixture was deoxygenated by several freeze–thaw cycles. The reaction vessel was then filled with argon and stirred next to a sunlamp for 3 min. After stirring in the absence of light for an additional 10 min the white precipitate that formed during the course of the reaction was removed by dilution with hexane and filtration through a pad of celite. The filtrate was then concentrated and rapidly passed through a short silica gel column to give the product (16 mg, 100%) as a pale yellow foam. Spectroscopic analysis indicated that a mixture of α- and β-diastereomers was present. This material was dissolved in tetrahydrofuran (0.7 ml), cooled to 0°C, and treated with a solution of sodium periodate (26 mg, 0.12 mmol) in water (0.3 ml). After standing for 12 h at room temperature, the reaction mixture was diluted with ether, washed with water and brine, dried (MgSO₄) and concentrated. Column chromatography on silica gel (elution with 15% ethyl acetate–hexane) gave the pure olefin (7.7 mg, 89%) as a colourless oil that crystallized on standing.

From ref. [163] with permission.

General procedure for photostimulated reactions of mercurials with diaryl dichalcogenides

The mercurial and the coreactant were dissolved in 10 ml of a deoxygenated solvent under a nitrogen atmosphere in a Pyrex flask equipped with a magnetic stirring bar and a rubber septum. The stirred mixture was irradiated by either a 275 W sunlamp about 20 cm from the flask or in a 350 nm Rayonet photoreactor. After completion of the reaction, the mixture was decanted from mercury metal or filtered from the precipitate of mercury salt. Benzene was removed under vacuum, and the residue was analysed by ¹H NMR, GLC and GCMS. When Me₂SO was used as the solvent, the reaction mixture was poured into water, and the product was extracted with benzene. The extract was washed twice with water, dried over anhydrous sodium sulphate and concentrated. The residue was then analysed by ¹H NMR, GLC and GCMS.

From ref. [156] with permission.

3.8.3.2 *From organocobalt compounds*

Like their organomercury counterparts in the preceding section, a variety of organocobalt complexes undergo smooth bimolecular homolytic substitution reactions with disulphides and diselenides. Once again, these reactions may be performed either thermally or photochemically, although the latter mode seems to be preferred.

Although earlier studies tended to focus on the cobaloxime system [165, 166] (e.g. dmgH Section 3.8.1), the Pattenden group [167] tend to recommend complexes based on salen or salophen ligands for their functional group interconversions.

The simplicity of the overall transformation is appropriately reflected in the typical example below and in some selected examples collated in Table 3.34. The use of acyl cobalt complexes is also possible [168]. This provides a useful route to thio- and selenoesters, although decarbonylation of the intermediate acyl radical may occur if benzylic or allylic radicals are formed.

$$R \cdot \longrightarrow R—X—Ph$$

$$R=ArCH_2 \cdot,$$

$$X = S, Se$$

TABLE 3.34

Preparation of alkyl phenyl sulphides and selenides from organocobalt complexes $(RCo(III)L)$ by photolysis in the presence of dichalcogenides $(Ph_2X_2, X = S, Se)$

Organocobalt complex		Product (RXPh)		
R	L	X	Yield (%)	Ref.
(benzofuranylmethyl)	(salen) (py)	S	85	[168]
	(salen) (py)	Se	75	[168]
$CH_3(CH_2)_9$	$(dmgH)_2 (py)$	S	82	[169]
$CH_3(CH_2)_8—CH(CH_3)$	$(dmgH)_2 (py)$	Se	98	[169]
(allyl)	$(dmgH)_2 (py)$	Se	41	[165]
$CH_3CH_2—C(=O)$	(salen) (py)	CH_3CH_2COSPh	70	[170]
$PhCH_2—C(=O)$	(salen) (py)	$PhCH_2SPh$	70	[170]

General photolytic procedure for the preparation of decyl phenyl sulphide

$$CH_3(CH_2)_9Co(III)(dmgH)_2(py) \xrightarrow[\text{PhSSPh}]{hv} CH_3(CH_2)_9SPh$$

Photolyses were performed in Pyrex tubes equipped with ground glass joints and fitted with rubber septa under a positive pressure of deoxygenated argon or nitrogen (syringe needle through the septum) in magnetically stirred argon or nitrogen saturated solutions, prepared by bubbling the gas through the solutions for 10–15 min. Solvent or reagent additions were performed with syringes or cannulas by standard anaerobic transfer techniques. Each reaction tube was immersed in a beaker of water that was cooled with cold tap water running through a coil of copper tubing immersed in the beaker surrounding the reaction tube. The water in the beaker was magnetically stirred along with the reactions. The light source was either a Sylvania 300 W incandescent lamp or a GE 500 W incandescent lamp, mounted in a ceramic socket. The end of the light source was placed 8–15 cm from the outside of the beaker. The entire apparatus was wrapped in foil during the photolysis, and the light bulb was cooled with a stream of air blown over the bulb from behind. Silica gel TLC for $RCo(III)(dmgH)_2py$ disappearance was performed by removing small samples of the reaction mixtures by a long needle inserted through the septum. Whenever a crude reaction mixture was filtered through silica gel prior to ${}^1H$ NMR or GC analysis, petroleum ether or methylene chloride was used as the solvent for eluting the desired product(s) from the silica gel without eluting the polar by-products.

The organocobalt complex (9.9 mmol) and diphenyl disulphide (14.8 mmol) were placed into 20 ml of benzene in a 25 ml Pyrex tube. The tube was sealed with a rubber septum and the solution deoxygenated with nitrogen. The reaction mixture was then photolysed for 180 min as described in the previous section. The tube was opened, the contents transferred to a round bottomed one necked flask and the solvent was removed on a rotary evaporator. The crude product was purified by silica gel chromatography. Yields were determined by ${}^1H$ NMR using triphenylmethane as the internal standard.

From ref. [169] with permission.

REFERENCES

1. J. G. Noltes and G. J. M. Van der Kerk, *Chem. Ind.* 294 (1959).
2. H. G. Kuivila, *Synthesis* 499 (1970); H. G. Kuivila, *Adv. Organometal. Chem.* **1**, 47 (1964).
3. W. P. Neumann, *Synthesis* 665 (1987).
4. M. Pereyre, J.-P. Quintard and A. Rahm, *Tin in Organic Synthesis*. Butterworths, London, 1987.
5. E. C. Friedrich and R. L. Holmstead, *J. Org. Chem.* **36**, 971 (1971); E. C. Friedrich and R. L. Holmstead, *J. Org. Chem.* **37**, 2546 (1972).

6. L. J. Altman and R. C. Baldwin, *Tetrahedron Lett.* 2531 (1971).
7. J. A. Aimetti, E. S. Hamanaka, D. A. Johnson and M. S. Kellogg, *Tetrahedron Lett.*, 4631 (1979).
8. H. Laurent and R. Wiechert, German Patent 2320999 (1974); *Chem. Abstracts* **82**, 43657 (1975).
9. B. Giese and J. Dupuis, *Tetrahedron Lett.* **25** 1349 (1984).
10. J. P. Praly, *Tetrahedron Lett.* **24**, 3075 (1983).
11. K. Annen, H. Hofmeister, H. Laurent and R. Wiechert, *Liebigs Ann. Chem.* 705 (1983).
12. C. R. Engel and D. Mukherjee, *Steroids*, **37**, 73 (1981).
13. Z. J. Duri, B. M. Fraga and J. R. Hanson, *J. Chem. Soc. Perkin Trans. I*, 161 (1981).
14. G. L. Grady, *Synthesis*, 255 (1971).
15. G. L. Grady and S. K. Chokshi, *Synthesis*, 483 (1972).
16. A. L. J. Beckwith and S. W. Westwood, *Aust. J. Chem.* **36**, 2123 (1983); D. P. G. Hamon and K. R. Richards, *Aust. J. Chem.* **36**, 109 and 2243 (1983).
17. D. R. Williams, B. A. Barner, K. Nishitani and J. G. Phillips, *J. Am. Chem. Soc.* **104**, 4708 (1982).
18. H. O. House, S. G. Boots and V. K. Jones, *J. Org. Chem.* **30**, 2519 (1965).
19. A. J. Bloodworth and H. J. Eggelte, *J. Chem. Soc. Chem. Commun.* 865 (1982).
20. K. E. Coblens, V. B. Muralidharan and B. Ganem, *J. Org. Chem.* **47**, 5041 (1982).
21. G. R. Krow, D. A. Shaw, C. S. Jovais and H. G. Ramjit, *Synth. Commun.* 575 (1983).
22. S. Ogawa, T. Ueda, Y. Funaki, Y. Hongo, A. Kasuga and T. Suami, *J. Org. Chem.* **42**, 3083 (1977).
23. N. Hong, M. Funabashi and J. Yoshimura, *Carbohydrate Res.* **96**, 21 (1981).
24. H. Redlich and W. Roy, *Liebigs Ann. Chem.* 1215 (1981).
25. P. G. Sammes and D. Thetford, *J. Chem. Soc. Perkin Trans. I*, 111 (1988).
26. H. Hřebabecký, J. Brokeš and J. Beránek, *Coll. Czech. Chem. Commun.* 599 (1980).
27. D. G. Norman and C. B. Reese, *Synthesis* 304 (1983).
28. W. Boland and L. Jaenicke, *Chem. Ber.* **110**, 1823 (1977).
29. Y. V. Quang, D. Carniato, L. V. Quang and F. Le Goffic, *Synthesis* 62 (1985).
30. J. T. Groves and S. Kittisopikul, *Tetrahedron Lett.* 4291 (1977).
31. S. Takano, S. Nishizawa, M. Akiyama and K. Ogasawara, *Synthesis* 949 (1984).
32. L. Fitjer and J. M. Conia, *Angew. Chem. Int. Ed. Engl.* **12**, 332 (1973).
33. W. P. Neumann and H. Hillgärtner, *Synthesis* 537 (1971).
34. J. Vihanto, *Acta. Chem. Scand. Ser. B.* **B 37**, 703 (1983).
35. A. Medici, M. Fogagnolo, P. Pedrini and A. Dondoni, *J. Org. Chem.* **47**, 3844 (1982).
36. Review: W. Hartwig, *Tetrahedron* **39**, 2609 (1983).
37. D. H. R. Barton and S. W. McCombie, *J. Chem. Soc. Perkin Trans. I*, 1574 (1975), 37a; J. J. Fox, N. Miller and I. Wempen, *J. Med. Chem.* **9**, 101 (1966).
38. D. H. R. Barton and W. B. Motherwell, New and selective reactions and reagents, in *Organic Synthesis Today and Tomorrow* (IUPAC) (ed. B. M. Trost and C. R. Hutchinson). Pergamon Press, Oxford, 1981; D. H. R. Barton and W. B. Motherwell, *Pure Appl. Chem.* **53**, 15 (1981).
39. M. J. Robins, J. S. Wilson and F. Hansske, *J. Am. Chem. Soc.* **105**, 4059 (1983).
40. E. J. Prisbe and J. C. Martin, *Synth. Commun.* **15**, 401 (1985).
41. D. Crich, Unpublished results.
42. S. Kim and K. Y. Ki, *J. Org. Chem.* **51**, 2613 (1986).
43. D. H. R. Barton, D. Crich, A. Löberding and S. Z. Zard, *Tetrahedron* **42**, 2329 (1986); P. J. Barker and A. L. J. Beckwith, *J. Chem. Soc. Chem. Commun.* 683 (1984); M. D. Bachi and E. Bosch, *J. Chem. Soc. Perkin Trans. I*, 1517 (1988); D. Crich, *Tetrahedron Lett.* **29**, 5805 (1988); J. E. Forbes and S. Z. Zard, *Tetrahedron Lett.* **30**, 4367 (1989).

44. D. H. R. Barton, W. B. Motherwell and A. Stange, *Synthesis* 743 (1981).
45. D. H. R. Barton, W. Hartwig and W. B. Motherwell, *J. Chem. Soc. Chem. Commun.* 447 (1982).
46. E. Zbiral, H. H. Brandstetter and E. P. Schreiner, *Monatsh. Chem.* **119**, 127 (1988).
47. K. Tatsuta, K. Akimoto and M. Kinoshita, *J. Am. Chem. Soc.* **101**, 6116 (1979).
48. R. E. Carney, J. B. McAlpine, M. Jackson, R. S. Stanaszek, W. H. Washburn, M. Cirovic and S. L. Mueller, *J. Antibiotics* **31**, 441 (1978).
49. C. Copeland and R. V. Stick, *Aust. J. Chem.* **30**, 1269 (1977); J. J. Patroni and R. V. Stick, *Aust. J. Chem.* **31**, 445 (1978); J. J. Patroni and R. V. Stick, *J. Chem. Soc. Chem. Commun.* 449 (1978); J. J. Patroni and R. V. Stick, *Aust. J. Chem.*. **32**, 411 (1979).
50. S. Iacono and J. R. Rasmussen, *Org. Synth.* **64**, 57 (1986).
51. H. Hagiwara and H. Uda, *J. Chem. Soc. Chem. Commun.* 815 (1988).
52. Ohta Pharmaceutical Co. Ltd, Japan, Kokai Tokkyo Koho, JP 59 16852 (1984); *Chem. Abs.* **101**, 90656 (1984).
53. M. H. Beale, P. Gaskin. P. S. Kirkwood and J. MacMillan, *J. Chem. Soc. Perkin Trans. I*, 885 (1980).
54. D. B. Tulshian and B. Fraser-Reid, *Tetrahedron Lett.* **21**, 4549 (1980).
55. D. J. Hart, *J. Org. Chem.* **46**, 367 (1981).
56. T. Rosen, M. J. Taschner and C. H. Heathcock, *J. Org. Chem.* **49**, 3994 (1984).
57. J. R. Rasmussen, *J. Org. Chem.* **45**, 2725 (1980).
58. J. Defaye, H. Driguez, B. Henrissat and E. Bar-Guilloux, *Nouv. J. Chim.* **4**, 59 (1980).
59. T. Hayashi, T. Iwaoka, N. Takeda and E. Ohki, *Chem. Pharm. Bull.* **26**, 1786 (1978).
60. D. H. R. Barton, W. Hartwig, R. S. Hay-Motherwell, W. B. Motherwell and A. Stange, *Tetrahedron Lett.* **23**, 2019 (1982).
61. D. H. R. Barton and D. Crich, *J. Chem. Soc. Perkin Trans. I*, 1603 (1986).
62. D. H. R. Barton and R. Subramanian, *J. Chem. Soc. Perkin Trans. I*, 1718 (1977).
62a. F. E. Ziegler and Z. Zheng, *Tetrahedron Lett.* **28**, 5973 (1987).
63. J. P. Kutney, T. Honda, A. V. Joshua, N. G. Lewis and B. R. Worth, *Helv. Chim. Acta* **61**, 690 (1978).
64. Y. Ueno, C. Tanaka and M. Okawara, *Chem. Lett.* 795 (1983).
65. H. Sano, T. Takeda and T. Migita, *Chem. Lett.* 119 (1988).
66. H. Redlich, H.-J. Neumann and H. Paulsen, *Chem. Ber.* **110**, 2911 (1977).
67. S. C. Dolan and J. MacMillan, *J. Chem. Soc. Chem. Commun.* 1588 (1985).
68. A. Chu and L. N. Mander, *Tetrahedron Lett.* **29**, 2727 (1988).
69. R. A. Jackson and F. Malek, *J. Chem. Soc. Perkin Trans. I*, 1207 (1980).
70. J. Pfenninger, C. Heuberger and W. Graf, *Helv. Chim. Acta* **63**, 2328 (1980).
71. T. Saegusa, S. Kobayashi, Y. Ito and N. Yasuda, *J. Am. Chem. Soc.* **90**, 4182 (1968).
72. D. H. R. Barton, G. Bringmann, G. Lamotte, W. B. Motherwell, R. S. Hay-Motherwell and A. E. A. Porter, *J. Chem. Soc. Perkin Trans. I*, 2657 (1980).
73. D. H. R. Barton, G. Bringmann and W. B. Motherwell, *J. Chem. Soc. Perkin Trans. I*, 2665 (1980).
74. M. Philipe, B. Quiclet-Sire, A. M. Sepulchre, S. D. Gero, H. Loibner, W. Streicher, P. Stuetz and N. Moreau, *J. Antibiotics* **36**, 250 (1983).
75. D. I. John, N. D. Tyrrell and E. J. Thomas, *Tetrahedron* **39**, 2477 (1983).
76. D. I. John, N. D. Tyrrell and E. J. Thomas, *J. Chem. Soc. Chem. Commun.* 901 (1981).
77. D. H. R. Barton, G. Bringmann and W. B. Motherwell, *Synthesis* 68 (1980).
78. Review: N. Ono and A. Kaji, *Synthesis* 693 (1986).
79. N. Ono, R. Tamura, H. Miyake and A. Kaji, *Tetrahedron Lett.* **22**, 1705 (1981).
80. D. D. Tanner, E. V. Blackburn and G. E. Diaz, *J. Am. Chem. Soc.* **103**, 1557 (1981); J.

Dupuis, B. Giese, J. Hartung, M. Leising, H. G. Korth and R. Sustmann, *J. Am. Chem. Soc.* **107**, 4332 (1985).

81. N. Ono, H. Miyake, A. Kamimura, I. Hamamoto, R. Tamura and A. Kaji, *Tetrahedron* **41**, 4013 (1985).
82. N. Ono, H. Miyake and A. Kaji, *J. Org. Chem.* **49**, 4997 (1984).
83. N. Ono, A. Kamimura, H. Miyake, I. Hamamoto and A. Kaji, *J. Org. Chem.* **50**, 3692 (1985).
84. G. Rosini, R. Ballini and M. Petrini, *Synthesis* 269 (1985).
85. G. Rosini, M. Petrini, R. Ballini and P. Sorrenti, *Tetrahedron* **40**, 3809 (1984).
86. N. Ono, I. Hamamoto, H. Miyake and A. Kaji, *Chem. Lett.* 1079 (1982).
87. F. Baumberger and A. Vasella, *Helv. Chim. Acta* **66**, 2210 (1983).
88. N. Ono, H. Miyake, A. Kamimura and A. Kaji, *J. Chem. Soc. Perkin Trans. I*, 1929 (1987).
89. D. Seebach and P. Knochel, *Helv. Chim. Acta* **67**, 261 (1984).
90. T. Tanaka, T. Toru, N. Okamura, A. Hazato, S. Sugiura, K. Manabe, S. Kurozumi, M. Suzuki, T. Kawagishi and R. Noyori, *Tetrahedron Lett.* 4103 (1983).
91. N. Ono, H. Miyake and A. Kaji, *J. Chem. Soc. Chem. Commun.* 875 (1983).
92. R. Meuwly and A. Vasella, *Helv. Chim. Acta* **68**, 997 (1985).
93. For a relative rates of cleavage see: A. L. J. Beckwith and P. E. Pigou, *Aust. J. Chem.* **39**, 77 (1986).
94. F. W. Hoffmann, R. J. Ess, T. C. Simmons and R. S. Hanzel, *J. Am. Chem. Soc.* **78**, 6414 (1956).
95. C. Walling and R. Rabinowitz, *J. Am. Chem. Soc.* **79**, 5326 (1957); C. Walling and R. Rabinowitz, *J. Am. Chem. Soc.* **81**, 1243 (1959).
96. M. Pang and E. I. Becker, *J. Org. Chem.* **29**, 1948 (1964).
97. E. Vedejs and D. W. Powell, *J. Am. Chem. Soc.* **104**, 2046 (1982).
98. G. A. Krafft and P. T. Meinke, *Tetrahedron Lett.* **26**, 135 (1985).
99. J. M. McIntosh and C. K. Schram, *Can. J. Chem.* **55**, 3755 (1977).
100. H. Natsugari, Y. Matsushita, N. Tamura, K. Yoshioka and M. Ochiai, *J. Chem. Soc. Perkin Trans. I*, 403 (1983).
101. J. D. Buynak, M. N. Rao, H. Pajouhesh, R. Y. Chandrasekaran, K. Finn, P. de Meester and S. C. Chu, *J. Org. Chem.* **50**, 4245 (1985).
102. G. A. Kraus and J. O. Nagy, *Tetrahedron Lett.* **24**, 3427 (1983).
103. D. Kahne, D. Yang, J. J. Lim, R. Miller and E. Papuaga, *J. Am. Chem. Soc.* **110**, 8716 (1988).
104. T. H. Haskell, P. W. K. Woo and D. R. Watson, *J. Org. Chem.* **42**, 1302 (1977).
105. C. G. Gutierrez, R. A. Stringham. T. Nitasaka and K. G. Glasscock, *J. Org. Chem.* **45**, 3393 (1980).
106. C. G. Gutierrez, C. P. Alexis and J. M. Uribe, *J. Chem. Soc. Chem. Commun.* 124 (1984).
107. D. R. Williams and J. L. Moore, *Tetrahedron Lett.* **24**, 339 (1983).
108. Review and book: T. G. Back, Radical reactions of selenium compounds, in *Organoselenium Chemistry* (ed. D. Liotta). Wiley, New York, 1987.
109. D. L. J. Clive, G. J. Chittatu, V. Farina, W. A. Kiel, S. M. Menchen, C. G. Russell, A. Singh, C. K. Wong and N. J. Curtis, *J. Am. Chem. Soc.* **102**, 4438 (1980).
110. P. J. Giddings, D. I. John and E. J. Thomas, *Tetrahedron Lett.* **21**, 399 (1980); For a related example see K. Hirai, Y. Iwano and K. Fujimoto, *Tetrahedron Lett.* **23**, 4021 (1982).
111. D. L. J. Clive, G. Chittatu and C. K. Wong, *J. Chem. Soc. Chem. Commun.* 41 (1978).
112. J. N. Denis and A. Krief, *Tetrahedron Lett.* **23**, 3411 (1982).
113. K. C. Nicolau, S. P. Seitz, W. J. Sipio and J. F. Blount, *J. Am. Chem. Soc.* **101**, 3884 (1979); K. C. Nicolau, R. L. Magolda, W. J. Sipio, W. E. Barnette, Z. Lysenko and M. M. Joullie, *J. Am. Chem. Soc.* **102**, 3784 (1980).
114. J. Lucchetti and A. Krief, *Tetrahedron Lett.* **22**, 1623 (1981).

115. Reviews: D. H. R. Barton and W. B. Motherwell, *Heterocycles* **21**, 1 (1984); D. H. R. Barton and S. Z. Zard, *Pure Appl. Chem.* **58**, 675 (1986); D. Crich, *Aldrichimica Acta* **20**, 35 (1987); D. H. R. Barton and S. Z. Zard, *Janssen Chimica Acta* **4**, 3 (1987); D. H. R. Barton and N. Ozbalik, *Paramagnetic Organomet. Species Activation, Selectivity, Catalysis* 1 (1989).
116. D. H. R. Barton, H. A. Dowlatshahi, W. B. Motherwell and D. Villemin, *J. Chem. Soc. Chem. Commun.* 732 (1980).
117. D. H. R. Barton, D. Crich and W. B. Motherwell, *J. Chem. Soc. Chem. Commun.* 939 (1983).
118. D. H. R. Barton, D. Crich and W. B. Motherwell, *Tetrahedron* **41**, 3901 (1985).
119. D. H. R. Barton, Y. Hervé, P. Potier and J. Thierry, *J. Chem. Soc. Chem. Commun.* 1298 (1984); D. H. R. Barton, Y. Hervé, P. Potier and J. Thierry, *Tetrahedron* **44**, 5479 (1988).
120. E. W. Della and J. Tsanaktsidis, *Aus. J. Chem.* **39**, 2061 (1986).
121. M. Ihara, M. Suzuki, K. Fukumoto, T. Kametani and C. Kabuto, *J. Am. Chem. Soc.* **110**, 1963 (1988).
122. D. Crich and T. J. Ritchie, *J. Chem. Soc. Chem. Commun.* 1461 (1988).
123. For additional examples involving reductive decarboxylation using tertiary mercaptans see *inter alia*: J. C. Braeckman, D. Daloze, M. Kaisin and B. Moussiaux, *Tetrahedron* **41**, 4603 (1985); O. Campopiano, R. D. Little and J. L. Petersen, *J. Am. Chem. Soc.* **107**, 3721 (1985); A. Otterbach and H. Musso, *Angew. Chem. Int. Ed. Engl.* **26**, 554 (1987); J. D. Winkler and V. Sridar, *J. Am. Chem. Soc.* **108**, 1708 (1986); J. D. Winkler, J. P. Hey and P. G. Williard, *J. Am. Chem. Soc.* **108**, 6425 (1986); J. D. Winkler, K. F. Heuegar and P. G. Williard, *J. Am. Chem. Soc.* **109**, 2850 (1987).
124. D. H. R. Barton, D. Crich and P. Potier, *Tetrahedron Lett.* **26**, 5943 (1985).
125. D. H. R. Barton, D. Crich and G. Kretzschmar, *J. Chem. Soc. Perkin Trans. I*, 39 (1986).
126. D. H. R. Barton, D. Crich and W. B. Motherwell, *Tetrahedron Lett.* **24**, 4979 (1983).
127. Professor S. Ikegami, Private communication to Professor Sir Derek Barton.
128. G. W. J. Fleet, J. C. Son, J. M. Peach and T. A. Hamor, *Tetrahedron Lett.* **29**, 1449 (1988).
129. For other examples see *inter alia*: L. Rosslein and C. Tamm, *Helv. Chim. Acta* **71**, 47 (1988); K. Kamiyama, S. Kobayashi and M. Ohno, *Chem. Lett.* 29 (1987).
130. E. Vogel, T. Schieb, W. H. Schulz, K. Schmidt, H. Schmickler and J. Lex, *Angew. Chem. Int. Ed. Engl.* **25**, 723 (1986).
131. D. H. R. Barton, B. Lacher and S. Z. Zard, *Tetrahedron Lett.* **26**, 5939 (1985); D. H. R. Barton, B. Lacher and S. Z. Zard, *Tetrahedron* **43**, 4321 (1987).
132. D. H. R. Barton, J. Boivin, D. Crich and C. H. Hill, *J. Chem. Soc. Perkin Trans. I*, 1805 (1986).
133. D. H. R. Barton, D. Bridon and S. Z. Zard, *Tetrahedron Lett.* **25**, 5777 (1984).
134. D. H. R. Barton, D. Bridon and S. Z. Zard, *Heterocycles* **25**, 449 (1987).
135. D. H. R. Barton, D. Bridon, Y. Hervé, P. Potier, J. Thierry and S. Z. Zard, *Tetrahedron* **42**, 4983 (1986).
136. D. H. R. Barton, D. Bridon and S. Z. Zard, *Tetrahedron Lett.* **27**, 4309 (1986).
137. D. H. R. Barton, B. Lacher, B. Misterkiewicz and S. Z. Zard, *Tetrahedron* **44**, 1153 (1988).
138. D. H. R. Barton, D. Crich and W. B. Motherwell, *J. Chem. Soc. Chem. Commun.* 242 (1984).
139. D. H. R. Barton, D. Bridon and S. Z. Zard, *J. Chem. Soc. Chem. Commun* 1066 (1985).
140. For useful reviews on radical reactions of organoboron compounds see *inter alia*: A. Pelter and K. Smith, in *Comprehensive Organic Chemistry* (ed. D. H. R. Barton and W. D. Ollis), Vol. 3, Sect. 14.3, p. 791. Pergamon Press, Oxford, 1979; H. C. Brown and M. M. Midland, *Angew. Chem. Int. Ed. Engl.* **11**, 692 (1972).
141. For useful reviews on radical reactions of organomercury compounds see *inter alia*: R. C. Larock, *Solvomercuration/Demercuration Reactions in Organic Synthesis*. Springer-Verlag,

Berlin, 1986; G. A. Russell, *Acc. Chem. Res.* **22**, 1 (1989); J. Barluenga and M. Yus, *Chem. Rev.* **88**, 487 (1988).

142. H. C. Brown and G. W. Kabalka, *J. Am. Chem. Soc.* **92**, 712, 714 (1970).
143. For useful reviews on radical aspects of organocobalt chemistry see *inter alia*: D. Dodd and M. D. Johnson, *J. Organomet. Chem.* **52**, 1 (1973); R. Scheffold, G. Rytz and L. Walder, *Modern Synth. Methods* **3**, 355 (1983); G. Pattenden, *Chem. Soc. Rev.* **17**, 361 (1988).
144. G. N. Schrauzer, *Angew. Chem. Int. Ed. Engl.* **15**, 417 (1976); E. G. Samsel and J. K. Kochi, *J. Am. Chem. Soc.* **108**, 4790 (1986).
145. A. G. Davies and B. P. Roberts, *J. Chem. Soc.* B 311 (1969); H. C. Brown, M. M. Midland and G. W. Kabalka, *J. Am. Chem. Soc.* **93**, 1024 (1971).
146. H. C. Brown, M. M. Midland and G. W. Kabalka, *Tetrahedron* **42**, 5523 (1986).
147. C. L. Hill and G. M. Whitesides, *J. Am. Chem. Soc.* **96**, 870 (1974).
148. R. P. Quirk and R. E. Lea, *J. Am. Chem. Soc.* **98**, 5973 (1976).
149. J. C. Sih and D. R. Graber, *J. Org. Chem.* **47**, 4919 (1982).
150. R. C. Bernotas and B. Ganem, *Tetrahedron Lett.* **26**, 1123 (1985).
151. K. E. Harding, T. H. Marman and D.-H. Nam, *Tetrahedron* **44**, 5605 (1988).
152. P. J. Toscano and L. G. Marzilli, *Progress Inorg. Chem.* **31**, 105 (1984).
153. P. J. Toscano and L. G. Marzilli, *Progress. Inorg. Chem.* **31**, 155 (1984).
154. D. J. Coveney, V. F. Patel and G. Pattenden, *Tetrahedron Lett.* **28**, 5949 (1987).
155. T. Okamoto and S. Oka, *J. Org. Chem.* **49**, 1589 (1984).
156. G. A. Russell, P. Ngoviwatchai, H. I. Tashtoush, A. Pla-Dalmau and R. K. Khanna, *J. Am. Chem. Soc.* **110**, 3530 (1988).
157. G. A. Russell and H. I. Tashtoush, *J. Am. Chem. Soc.* **105**, 1398 (1983).
158. G. A. Russell, *Acc. Chem. Res.* **22**, 1 (1989).
159. T. Toru, Y. Yamada, E. Maekawa and Y. Ueno, *Chem. Lett.* 1827 (1987).
160. G. A. Russell and J. Hershberger, *J. Am. Chem. Soc.* **102**, 7603 (1980).
161. G. A. Russell, P. Ngoviwatchai, H. Tashtoush and J. Hershberger, *Organometallics* **6**, 1414 (1987).
162. J. E. McMurry and M. D. Erion, *J. Am. Chem. Soc.* **107**, 2712 (1985).
163. M. D. Erion and J. E. McMurry, *Tetrahedron Lett.* **26**, 559 (1985).
164. G. A. Russell and P. Ngoviwatchai, *Tetrahedron Lett.* **26**, 4975 (1985); G. A. Russell and P. Ngoviwatchai, *Tetrahedron Lett.* **27**, 3479 (1986); see also ref. [160].
165. J. Deniau, K. N. V. Duong, A. Gaudemar, P. Bougeard and M. D. Johnson, *J. Chem. Soc. Perkin Trans. II*, 393 (1981).
166. J. Deniau, K. N. V. Duong, C. Merienne and A. Gaudemar, *Bull. Soc. Chim. Fr. II*, 180 (1983).
167. G. Pattenden, *Chem. Soc. Rev.* **17**, 361 (1988).
168. D. J. Coveney, V. F. Patel and G. Pattenden, *Tetrahedron Lett.* **28**, 5949 (1987).
169. B. P. Branchaud, M. S. Meier and M. N. Malekzadeh, *J. Org. Chem.* **52**, 212 (1987).
170. V. F. Patel and G. Pattenden, *Tetrahedron Lett.* **29**, 707 (1988).

–4–

Olefin Forming β-Elimination Reactions

4.1 INTRODUCTION

The dangers of competing elimination reactions by generation of a carbanion adjacent to a leaving group such as a halogen or a mesylate are well known in traditional ionic synthesis. Within the last decade, an analogous series of radical leaving groups has evolved. Casual inspection of their general nature reveals the complimentarity of the ionic and free radical processes.

$$X = Br, \quad O \overset{S}{\underset{}{\diagup}} SMe, SPh, SOPh, SO_2Ph, SePh, NO_2$$
$$X \neq OAc, OSO_2CH_3$$

While such constraints may be regarded as limiting in terms of synthetic planning if an alternative reaction for the carbon centred radical is desired, olefin forming fragmentation reactions are nevertheless of preparative value in their own right.

In terms of stereospecific olefin synthesis in the acyclic series, some degree of control is possible, although the stereoelectronic preference for a *trans* coplanar arrangement found in ionic eliminations is generally much less dominant. This may occur, however, if the intermediate radical is capable of a predisposition towards a bridging structure with a preference for *anti* elimination (e.g. X = Br) or, more usefully, if the leaving group departs from a preferred conformer of a diastereoisomer faster than rotation of the intermediate radical can occur (e.g. X = SOPh) (*vide infra*).

4.2 ALKENES FROM VICINAL DIHALIDES

Historically, the regeneration of an olefin from a 1,2-dibromide was one of the earliest reactions to be investigated using tri-n-butyltin hydride [1].

Modest stereoselectivity in acyclic systems has been observed by operating at low temperature and relatively high stannane concentration. Thus, *meso*-2,3-dibromobutane gives a *trans–cis* ratio of 9.0, while the corresponding *dl* isomer affords a *cis–trans* ratio of 1.9 [1].

Vicinal dichlorides cannot, however, be used for double bond protection since elimination is a minor pathway and sequential replacement of halogen by hydrogen dominates.

The reduction of 2,3-dihalobutanes is representative for alkene formation from a vicinal dibromide [1].

Reduction of 2,3-dihalobutanes

Reduction of 2,3-dihalobutanes with organotin hydrides was carried out in a small, three necked reaction flask or in a Pyrex tube possessing a ground glass joint. The reactions were run in a constant temperature bath under a blanket of argon. The reactions of the vicinal dichlorides with tri-n-butyltin hydride and tri-n-butyltin deuteride and those of the vicinal dibromides with tri-n-butyltin hydride at $-75°C$ were photoinitiated with a 100 W General Electric mercury vapour lamp. A quartz tube (i.d. = 3.4 cm) containing the lamp was immersed in a constant temperature bath (22°C) or in a bath of methanol or petroleum ether cooled to $-78°C$. The tube containing the reactants was placed 3 cm from the lamp in the bath and 0.5 cm or less away from the lamp in the $-78°C$ bath.

A measured amount of the organotin hydride was introduced dropwise from a syringe through a rubber septum into the reaction flask containing the 2,3-dibromobutane. The reaction with dibromides began almost immediately upon contact of the reactants with each other. In reactions involving dichlorides, and those involving dibromides at $-75°C$, measured amounts of dihalide and tri-n-butyltin hydride or deuteride were introduced into a Pyrex tube under argon and then irradiated at the desired temperature until the reaction was complete.

The gaseous products were collected in a trap, cooled to $-78°C$, and analysed by GLPC at room temperature on a 15 foot (~ 4.5 m) column containing 28% dimethylsulpholane on C-22 Firebrick (40–60 mesh). The identities of the 2-butenes were ascertained on the basis of comparison of their infrared spectra and GLPC retention times with those of authentic samples. n-Butane was identified by comparison of its retention time with that of an authentic sample. Yields of 2-butenes were obtained by titration of the olefins with a standardized solution of bromine in acetic acid.

4.3 ALKENES FROM VICINAL DIOLS VIA 1,2-BIS-XANTHATE ESTERS

The transformation of a vicinal diol into an olefin *via* reaction of the derived *bis*-xanthate ester with tri-n-butylstannane has proven to be a particularly useful method in the carbohydrate area [2, 3].

In this instance, the lack of stereoselectivity in endocyclic double bond formation from an alicyclic pyranose may be considered as an advantage. Thus, as illustrated for the case of the rigid benzylidene derivative, good yields of olefin may be obtained either from the diequatorial *trans*-diol, or from its *cis* isomer.

Nevertheless, in acyclic systems some degree of control is possible in terms of channelling the product to the more thermodynamically stable alkene, as exemplified by the formation of *trans*-stilbene from both the *dl* and *meso* forms of hydrobenzoin.

TABLE 4.1

Alkene formation from vicinal dixanthates and tri-n-butylstannane [2, 3]

Substrate	Product	Yield (%)
		49
		62
		65

Additional examples of this reaction appear in Table 4.1 and a typical experimental procedure is given below.

Preparation of methyl-4,6-O-benzylidene-2,3-dideoxy-α-D-erythro-hex-2-enopyranoside

Tri-n-butylstannane (1.2 g) and the bis-dithiocarbonate (231 mg) in dry toluene (15 ml) were heated to reflux overnight, cooled, washed with aqueous potassium hydroxide, dried ($MgSO_4$) and evaporated. Chromatography on silica (diethyl ether–light petroleum gradient) and crystallization from ethanol gave the alkene (74 mg, 60%).

4.4 ALKENE FORMATION BY STANNANE INDUCED FRAGMENTATION OF THIOCARBONYL-O-ESTER DERIVATIVES OF β-HYDROXY SULPHIDES, SULPHONES AND ISOCYANIDES

By using a variety of thiocarbonyl-*O*-esters as precursors for formation of a carbon centred radical (see Section 3.2.3) the phenylthiyl and phenylsulphonyl radical leaving groups have also been brought into play in terms of preparatively useful elimination reactions.

$$X = SMe, \qquad Y = SPh$$
$$X = Ph, \qquad YSO_2Ph$$

4.4.1 Olefins from β-hydroxy sulphides [4, 5]

As in the case of vicinal dixanthates, both the *cis* and *trans* isomers of alicyclic substrates may be induced to give alkenes in a regiospecifically undemanding way, as shown in the accompanying examples [4].

The experimental procedure followed the method given by Barton and McCombie for simple deoxygenation (Section 3.2.3).

The leaving group ability of the phenylthiyl radical also features in a non-stereoselective aldehyde homologation sequence developed by Vatelle [5], and involves the synthesis of a precursor vinyl ether.

The examples studied, together with the isomeric ratio of enol ethers produced, are collected in Table 4.2. The preparation of 2-cyclohexyl-1-methoxyethene is typical.

TABLE 4.2

Preparation of vinyl ethers by tri-n-butylstannane induced fragmentation of 2-methoxy-2-phenylthio-1-xanthates [5]

Substrate	Product	
R–CH(OMe)CH(SPh)O–C(=S)SMe	$\begin{array}{c}R\\ \backslash C=C \diagup H\\ H \diagup \quad \backslash OMe\end{array}$	
	Yield (%)	Z:E
$R = $ benzyl (C₆H₅CH₂)	73	42:58
$R = $ cyclohexyl	73	40:60
$R = CH_3(CH_2)_8CH_2$	74	44:56
C_6H_5(phenyl)	82	33:67
$R = $ cinnamyl (C₆H₅CH=CHCH₂)	62	25:75
$R = $ (diene chain)	50	27:73

Typical procedure for the formation of vinyl ethers

Synthesis of S-methyl dithiocarbonates. To a solution of methoxyphenyl-thiomethyllithium in 10 ml of tetrahydrofuran, obtained by deprotonation of methoxymethyl phenyl thioether (1.51 g; 1.1 equiv.), with n-butyllithium at − 40°C for 45 min, was added, at − 78°C, 1 g of cyclohexane carboxaldehyde (8.9 mmole) in 10 ml of tetrahydrofuran. After 15 min at this temperature, 2.5 ml (5 equiv.) of carbon disulphide was added; after reaching room temperature and then refluxing for 30 min, methyl iodide (2.5 ml; 5 equiv.) was added, and refluxing continued for 30 min. After evaporation of volatiles, chromatography on silica gel of the residue (ether–hexane, 3:97) gave 2.55 g of a yellow liquid (80%) as a mixture of diastereoisomers.

Procedure for reductive elimination. To a solution of the xanthates described above (1.8 g; 5 mmole) in 10 ml of benzene were added 2.5 ml of tri-n-butylstannane (1.8 equiv.) and a catalytic amount of 2,2′-azobis(2-methylpropionitrile) (0.1 g). After refluxing the solution for 1 h (colourless solution), the solvent was evaporated. The residue was distilled (12 mmHg; 80–120°C) using a Kugelröhr apparatus and the distillate was further purified by chromatography on silica gel using ether–hexane–triethylamine (1:99:1) as the eluent to afford 0.541 g (77%) of 2-cyclohexyl-1-methoxyethene which had identical data to those reported in the literature.

From ref. [5] with permission.

4.4.2 Alkenes from β-hydroxy sulphones

The Julia olefination sequence has proved to be a cornerstone of modern organic synthesis. Problems sometimes arise, however, in the elimination step, which requires treatment of an ester of the initially formed β-hydroxy sulphone with sodium amalgam. An ingenious solution lay in the stannane mediated fragmentation of the thiocarbonyl-O-ester derivatives of the initial adducts. In contrast to the reactions of β-hydroxy sulphide derivatives, this route provides a highly stereoselective synthesis of *trans*-olefins, as illustrated by the preparation of *trans*-pentadec-7-ene, essentially free from the *cis* isomer [4, 6]

4.4.3 Alkene formation from a β-hydroxy isocyanide

An isolated example of olefin formation by "deoxydeamination" from the xanthate ester of a carbohydrate β-hydroxy isocyanide has been reported [7].

Interestingly, since 1,2-diisocyanides, like vicinal dichlorides (Section 4.2), do not tend to undergo elimination, the implication may be drawn that it is the isonitrile function which triggers elimination of a xanthyl residue and not vice versa.

4.5 OLEFIN FORMATION FROM β-HALOGENO SULPHIDES, SULPHOXIDES, SULPHONES AND SELENIDES

The generation of a carbon centred radical by dehalogenation using an organostannane provides an equally effective method for olefin generation provided that the neighbouring moiety is a good radical leaving group.

X = Cl, Br   Y = SPh, SePh, SO₂Ph, SOPh

In general terms for acyclic substrates, elimination of the phenylthio [4, 8] or phenylsulphonyl [4, 9] radical occurs in a non-stereospecific way, and the same mixture of isomeric olefins will be obtained from either the *erythro* or *threo* substrate. By way of contrast, however, the β-elimination of the phenylsulphinyl radical has been shown to occur at a faster rate than free rotation of the intermediate carbon centred radical with consequential implications for higher stereoselection [10]. Preparative applications of this observation could well be important if controlled diastereomer formation can be achieved. As in the case of thiocarbonyl ester triggered eliminations, however, the equal facility for olefin formation from either a *cis*- or *trans*-1,2-disubstituted alicyclic compound may be considered advantageous.

A selection of examples is shown in Table 4.3.

From the preparative standpoint, simple admixture of the substrate, tri-n-butylstannane (1.0–5.0 M equiv.) and AIBN (0.3 M equiv.) in benzene followed by heating at 80°C under an inert atmosphere is sufficient. The preparation of *cis*- and *trans*-but-2-ene from the bromo sulphones is typical [9] of the reaction conditions required.

TABLE 4.3

Alkene formation from β-halo sulphides, selenides, sulphoxides and sulphones using tri-n-butylstannane

Substrate	Product	E:Z ratio	Yield (%)	Ref.
		—	87	[4]
		2.31	60	[8]
		2.22	75	[8]
		—	—	[11]
		2.05	73	[9]
		1.99	66	[9]
		2.38	64	[10]
		0.89	71	[10]

TABLE 4.3 (*continued*)

Substrate	Product	$E:Z$ ratio	Yield (%)	Ref.
		0.469	69	[10]
		0.198	67	[10]

Olefin formation from β-bromosulphones

The appropriate diastereomeric bromosulphone (100 mg, 0.361 mmol), tri-n-butyl stannane (1.0–5.0 molar equiv.) and AIBN (59.3 mg, 0.361 mmol) in benzene (1.0 ml) were added, and the mixture was heated at 80°C. The gaseous products were continuously swept from the reactor through a water cooled spiral condenser into a trap at − 196°C by a stream of nitrogen. The contents of the trap were distilled under vacuum from − 78 to − 196°C. The distillate was analysed by infrared spectroscopy and gas chromatography. Gas chromatographic analysis of the benzene solution showed no detectable 2-(phenylsulphonyl)butane. Heating the sulphones under these reaction conditions did not result in interconversion of the diastereomers as determined by ^{1}H NMR.

4.6   ALKENES FROM VICINALLY FUNCTIONALIZED NITRO DERIVATIVES

At the present time, several examples of β-elimination featuring the nitro group have been described, in conjunction with nitro [12], sulphide [13] and sulphonyl [12] leaving groups.

In terms of stereoselectivity, the demonstrated capability of β-nitro sulphones [12] for *anti* elimination holds the greatest promise. This tendency is reflected to a lesser degree in the case of β-nitro sulphides and not at all in the reactions of vicinal dinitro compounds.

$$E : Z = 98 : 2$$

$$E : Z = 3 : 97$$

A possible explanation [13] for the observed stereochemical outcome is that a free alkyl radical intermediate is not involved and that fragmentation proceeds by a concerted process following addition of the tri-n-butylstannyl radical.

$$X = SO_2Ph, SPh$$

As in simple denitration reactions (Section 3.4) the versatility of the classical chemistry of the nitro group can be efficiently combined with reductive elimination. A typical example is the elegant method of the Ono group for the synthesis of allylic alcohols from nitroalkenes via Michael addition of the phenylthiolate anion and subsequent Henry reaction with formaldehyde [13].

TABLE 4.4

Preparation of alkenes from β-nitro sulphides, sulphones and vicinal dinitro compounds by reductive elimination with tri-n-butylstannane

Substrate	Product	E:Z ratio	Yield (%)	Ref.
(H, NO₂, Et, SPh, Me, OBz structure)	(H, OBz, Et, Me structure)	87:13	71	[13]
(Et, NO₂, H, PhS, Me, OBz structure)	(Et, OBz, H, Me structure)	27:73	71	[13]
(isoprenyl chain, PhS, NO₂, OH structure)	(dienol structure, OH)	—	93	[13]
(H, NO₂, OBz, SO₂Ph, Me structure)	(H, OBz, Me structure)	96:4	—	[12]
(NO₂, OBz, H, SO₂Ph, Me structure)	(H, OBz, Me structure)	12:88	—	[12]
(Me, Me, Et, NO₂, SO₂Ph, CO₂Et structure)	(Me, Me, Et, CO₂Et structure)	—	81	[12]
(lactone, SO₂Ph, Me, Me, NO₂ structure)	(lactone, isopropylidene structure)	—	72	[12]
(NO₂, Et, Me, Ph, Me, NO₂ structure) *erythro*	(Me, Et, Ph, Me structure)	1:1	75	[12]
(NO₂, Me, Me, Ph, Et, NO₂ structure) *threo*	(Me, Et, Ph, Me structure)	1:1	75	[12]
(bicyclopentyl dinitro structure, NO₂, NO₂)	(bicyclopentylidene structure)	—	72	[12]

Further examples of eliminations of this type are collected in Table 4.4, and the general "experimental method" which has been communicated is given below.

General procedure for the conversion of vicinal dinitro compounds or β-nitro sulphones to olefins [12]

The vicinal dinitro compound or β-nitro sulphone (1 equiv.) and tri-n-butyltin hydride (2.0 equiv.) were heated in benzene at 80°C for 1–2 h to give olefins in good yields. The reaction was somewhat accelerated by the presence of a small amount of AIBN. Pure olefins were isolated by distillation after column chromatography on silica gel to remove tin components.

Typical procedure for the conversion of nitroalkenes to homologous allylic alcohols

A mixture of 1-nitrocyclohexene (1.27 g, 10 mmol), thiophenol (1.36 g, 12.4 mmol), 37% formaldehyde (1.12 g, 13.1 mmol) and tetramethylguanidine (0.04 g, 0.34 mmol) in acetonitrile (5 ml) was stirred at room temperature for 24 h. The reaction mixture was poured into water, acidified with dilute hydrochloric acid and extracted with diethyl ether. The crude product obtained by the usual work up (washing with water, drying with anhydrous magnesium sulphate and removal of solvents) was subjected to column chromatography (silica gel/hexane–ethyl acetate) to give the β-nitro alcohol (2.34 g, 88% yield). Acetylation of the alcohol with acetic anhydride in pyridine then gave the corresponding acetate in quantitative yield. A mixture of the acetylated nitro sulphide (0.62 g, 2.0 mmol), tri-n-butylstannane (1.2 g, 4.1 mmol) and AIBN (0.08 g, 0.4 mmol) in benzene (4 ml) was heated at 80°C for 3 h. The crude product was purified by column chromatography (silica gel/benzene–hexane) to give 1-acetoxymethylcyclohexene (0.28 g, 91% yield).

From ref. [13] with permission.

REFERENCES

1. R. J. Strunk, P. M. DiGiacomo, K. Aso and H. G. Kuivila, *J. Am. Chem. Soc.* **92**, 2849 (1970).
2. A. G. M. Barrett, D. H. R. Barton, R. Bielski and S. W. McCombie, *J. Chem. Soc. Chem. Commun.* 866 (1977).

3. A. G. M. Barrett, D. H. R. Barton and R. Bielski, *J. Chem. Soc. Perkin Trans. I*, 2378 (1979).
4. B. Lythgoe and I. Waterhouse, *Tetrahedron Lett.* 4223 (1977).
5. J.-M. Vatele, *Tetrahedron Lett.* **25**, 5997 (1984).
6. P. J. Kocienski, B. Lythgoe and S. Ruston, *J. Chem. Soc. Perkin Trans. I*, 829 (1978).
7. D. H. R. Barton, G. Bringmann, G. Lamotte, W. B. Motherwell, R. S. Hay-Motherwell and A. E. A. Porter, *J. Chem. Soc. Perkin Trans. I*, 2657 (1980).
8. T. E. Boothe, J. L. Greene Jr. and P. B. Shevlin, *J. Am. Chem. Soc.* **98**, 951 (1976).
9. T. E. Boothe, J. L. Greene Jr. and P. B. Shevlin, *J. Org. Chem.* **45**, 794 (1980).
10. T. E. Boothe, J. L. Greene Jr., P. B. Shevlin, M. R. Willcott III, R. R. Inners and A. Cornelis, *J. Am. Chem. Soc.* **100**, 3874 (1978).
11. D. L. J. Clive, G. J. Chittattu, V. Farina, W. A. Kiel, S. M. Menchen, C. G. Russell, A. Singh, C. K. Wong and N. J. Curtis, *J. Am. Chem. Soc.* **102**, 4438 (1980).
12. N. Ono, H. Miyake, R. Tamura, I. Hamamoto and A. Kaji, *Chem. Lett.* 1139 (1981).
13. N. Ono, A. Kamimura and A. Kaji, *Tetrahedron Lett.* **25**, 5319 (1984).

—5—

Preparative Free Radical Rearrangement Reactions

5.1 INTRODUCTION

While direct functional group substitution and controlled olefin forming β-elimination reactions may now be considered as standard manipulations, recent preparative radical chain reactions involving rearrangement are, by comparison, relatively rare [1]. Indeed, the reluctance of carbon centred radicals to misbehave, either by participating in carbocationic like alkyl shifts, or by exhibiting unwanted basic character as in the case of some of their "nucleophilic" carbanion congeners, may be considered as one of their primary advantages. Nevertheless, a variety of potentially useful reactions has been observed over the years and are certainly worthy of further examination in the light of more modern methods for controlled radical generation.

Within the present chapter, some representative reactions which involve cleavage of a carbon–carbon or a carbon–oxygen bond by a neighbouring radical centre are discussed. The intramolecular addition of a carbon centred radical to a carbon–carbon multiple bond, as in cyclization of the 5-hexenyl radical, is exclusively discussed in Section 7.3.2, although it can also be formally considered as a rearrangement type.

From the preparative standpoint, it cannot be sufficiently re-emphasized that, in such reactions, the mode of addition of the reagents and their relative concentrations is absolutely crucial to success, and the desired product can only be predicted and obtained by application of sound kinetic knowledge of relative rates (cf. Chapter 1).

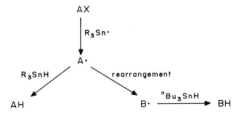

The stannane mediated reduction of a substrate AX, in which the intermediate carbon centred radial A* has a known propensity for rearrangement to a novel structure B*, may be considered by way of an example.

If the objective of the reaction is to prepare AH, then the correct procedure is to add the substrate AX to a neat or concentrated solution of a reactive stannane, e.g. R = Ph. Alternatively, if product BH is desired then slow addition of a poorer hydrogen atom donor such as tri-n-butylstannane to a dilute solution of AX is the favoured method. In general, many radical rearrangements are slow and therefore time must be given for them to occur.

5.2   ACETOXYL GROUP MIGRATION REACTIONS

An early study in the steroid series demonstrated the possibility of acetoxyl group migration with intermediate formation of a more stable tertiary radical from reduction of a secondary halide [2].

However, this particular class of rearrangement has been most extensively studied in the carbohydrate area, with particular emphasis on the nature of the carbon centred radical generated at the anomeric centre [3]. In the first instance, by operating at relatively high concentrations of stannane, hydrogen atom capture is effectively promoted, a neighbouring acetoxy group is unaffected and the "radical anomeric effect" operates to give high yields of axial product as demonstrated by deuterium labelling [4].

X = H,   Y = Cl   90 : 10 (100%)
X = Br, Cl, Y = H   90 : 10 (90%)

Slow addition of the stannane to the glycosyl halide, however, allows sufficient time for acetoxyl migration to occur prior to hydrogen atom capture and leads to an efficient synthesis of 2-deoxy sugar derivatives [5].

A variety of acetylated and benzoylated glycosyl bromides and iodides have been studied (Table 5.1) and the reaction has been shown to proceed in a highly stereoselective *cis* manner. A typical procedure follows.

TABLE 5.1

Preparation of 2-deoxy sugars from glycosyl halides by reductive rearrangement with tri-n-butylstannane [5]

Substrate	Product	Work up method	Yield (%)
		B	92
		B	71
		A	65
		B	70
		B	72

*General procedure for the preparation of 2-deoxy sugars
from glycosyl halides*

To a solution of tetra-*O*-acetylglycosyl bromide (0.02–0.05 mol) in benzene
or toluene at reflux a solution of tri-n-butyltin hydride (1.2 equiv.; 0.2–0.3 M)
and AIBN (0.2 equiv.) in benzene or toluene was added over 8–30 h via a
syringe fitted with a long needle. The apparatus was set up such that the needle
entered via the condenser and the tip was located 1 cm below the bottom
of the cooling jacket. After completion the reaction was cooled to 20°C and
the solvent evaporated. Work up was by method A or B.

Method A. The residue was taken up in acetonitrile (50 ml g^{-1}) and
extracted three times with an equal volume of hexane. Evaporation of the
acetonitrile fraction and crystallization from t-butyl methyl ether provided
the product.

Method B. The residue was taken up in ether–acetonitrile (9:1) and stirred
for 3 h at 20°C with potassium fluoride (3 g). The mixture was then filtered
on silica gel and the product isolated by evaporation of the solvent and
crystallization from t-butyl methyl ether.

From ref. [5] with permission.

A recent development in this area is the introduction of tris(trimethylsilyl)
silane by Chatgilialoglu [6]. Use of the poorer hydrogen atom donor is ideal
for rearrangement reactions by obviating the necessity for slow addition of
the stannane. A further significant practical advantage lies, of course, in the
production of silicon, as opposed to tin, by-products. In the case of the
present acetoxyl migration, this reaction can now be conducted as a simple
one pot operation. The following experimental details have been
communicated [7].

*Reduction of 2,3,4,6-tetraacetoxy-a-D-gluco- and galactopyranosyl
bromides with tris(trimethylsilyl)silane*

gluco,   71%
galacto, 70%

Direct mixing of the glycosyl halide, tris(trimethylsilyl) silane and AIBN, using 50 mmol solutions and a temperature of 80°C, led to complete reaction within 1 h.

5.3 RADICAL REARRANGEMENTS INVOLVING ALLYLIC SYSTEMS

In general terms, reduction of isomeric allylic substrates will lead to the same mixture of alkenes, provided that a common intermediate allylic radical is involved, as illustrated below for the simple case of butenyl halides [8]. Hence, in cases where regiospecific double bond placement is required allylic radicals should be avoided.

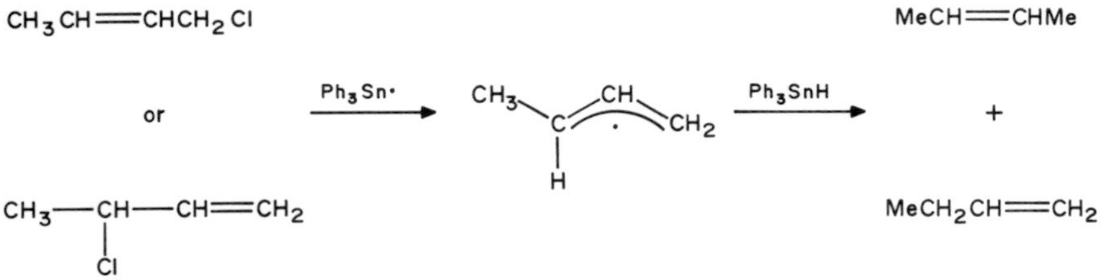

Nevertheless, an encouraging observation which hints at the possibility of stereoelectronically controlled reactions of allylic radicals may be found in the case of phenylsulphonyl substituted allylic halides [9].

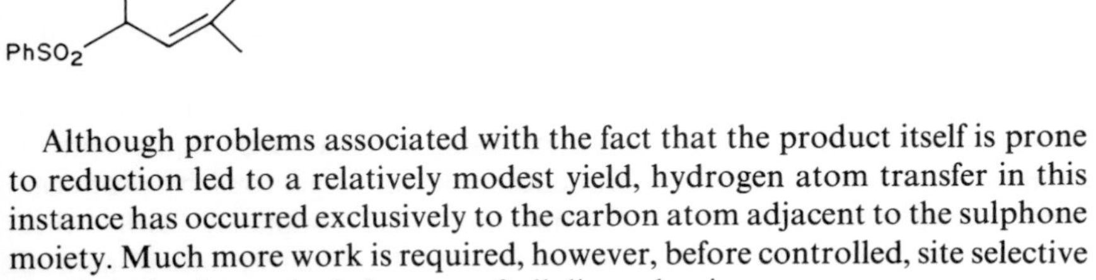

Although problems associated with the fact that the product itself is prone to reduction led to a relatively modest yield, hydrogen atom transfer in this instance has occurred exclusively to the carbon atom adjacent to the sulphone moiety. Much more work is required, however, before controlled, site selective reactions begin to rival the case of allylic carbanions.

Clean allylic functionalization reactions without the intermediacy of allylic radicals can, however, be achieved if a bimolecular S_H2' process or an addition–elimination sequence is involved. A particularly elegant example of this strategy may be found in a useful process for the regiospecific deoxygenation of primary allylic alcohols to their thermodynamically less stable terminal olefins [10].

This three step sequence involves the [3,3] sigmatropic rearrangement of an initially formed O-allyl xanthate ester followed by an S_H2' like reaction with tri-n-butylstannane. Protonolysis of the resultant allyl stannane then yields the terminal olefin. Careful monitoring of the reactions to give the rearranged dithiocarbonate is, of course, required in order to avoid regioisomeric contamination, since the reaction of allylic xanthates with tri-n-butyltin hydride is known to give the expected mixture of alkenes via the allylic radical [11]. Other advantages of the sequence include the facts that good yields are generally obtained using only a single equivalent of stannane and the stannylation sequence gives only gaseous by-products. Furthermore, the reaction of the product allyl stannanes with other electrophiles besides the proton may be used to generate functionalized alkenes. A typical procedure for this regiospecific deoxygenation follows, and additional examples appear in Table 5.2, including the last step in a total synthesis of ($\pm$)-rose oxide [12].

TABLE 5.2

Regiospecific deoxygenation of allylic alcohols via O-allyl xanthates

Allylic alcohol	Dithiocarbonate	Yield (%)	Allylic stannane	Yield (%)	Ref.
(structure) OH	(structure) SCOSMe	79	(structure) Sn^{n}Bu$_3$	78	[10]
(structure) OH	(structure) SCOSMe	64	(structure) Sn^{n}Bu$_3$	84	[10]
$^nC_9H_{19}$ (structure) OH	$^nC_9H_{19}$ (structure) SCOSMe	87	$^nC_9H_{19}$ (structure) Sn^{n}Bu$_3$	86	[10]
(structure) OH	(structure) SCOSMe	74	(structure) Sn^{n}Bu$_3$	91	[10]
(structure) OH	(structure) SCOSMe	56	(structure) Sn^{n}Bu$_3$	80	[10]
(structure) OH	(structure) SCOSMe	—	(structure)	64[a]	[12]

[a]Overall yield of product after protonolysis.

Regiospecific deoxygenation of allylic alcohols via O-allyl xanthates [10]

In situ formation of the O-allyl xanthate by sequential treatment of the allylic alcohol with sodium hydride, carbon disulphide and then methyl iodide (see Section 3.2.3 on deoxygenation) is followed by reflux in benzene for 2 h to give the rearranged dithiocarbonate.

Allyl transfer from sulphur to tin is then carried out by the following general procedure. Dithiocarbonate (1.8 mmol) and tri-n-butyltin hydride (3.6 mmol) in benzene (15 ml) containing catalytic amount of AIBN were heated at 80°C for about 2 h under nitrogen. After evaporation of benzene, the product was isolated by fractional distillation. It was found, however, that the employment of 1 equiv. of tri-n-butyltin hydride under similar conditions gave better results.

Reductive desulphonylation of allylic sulphones

The Ueno group have also studied the regioselective desulphonylation of allyl sulphones as a route to allyl stannanes [13]. As in the case of allylic dithiocarbonates a clean S_H2' process occurs with concomitant double bond migration.

The preparation of 1-(tri-n-butylstannyl)nona-2-ene, as in the above example, is typical of the procedure and additional examples are given in Table 5.3.

Regioselective desulphonylation of allylic sulphones—a typical example

A mixture of 3-tolyl-sulphonylnona-1-ene (447 mg, 1.38 mmol), tri-n-butyltin hydride (928 mg, 3.19 mmol) and AIBN (10 mg) in dry benzene (3 ml) was

TABLE 5.3

Regioselective desulphonylation of allyl sulphones with tri-n-butylstannane [13]

Sulphone	Product	Yield (%)
		68
		65
		74
		71

refluxed under a nitrogen atmosphere for 2h until the disappearance of the absorption of sulphone at 1320 and 1150 cm^{-1}. After completion of the reaction, 1-(tri-n-butylstannyl)nona-2-ene was isolated by column and neutral chromatography (neutral alumina, eluted with benzene) in 65% yield (443 mg) as a colourless oil. Further purification was carried out by Kugelrohr distillation under reduced pressure, b.p. 136–142°C (0.003 mmHg).

From ref. [13] with permission.

5.4 THE CYCLOPROPYLCARBINYL–HOMOALLYLIC RADICAL REARRANGEMENT

The rapid ring opening equilibrium between the cyclopropylcarbinyl radical and its open chain homoallylic counterpart is perhaps the most studied radical rearrangement to date. The rates involved have been accurately determined and used as a radical clock by Ingold and coworkers [14].

In the case of alkyl substituted derivatives, elegant studies of the equilibria involved have established that the manifold may be approached from either the homoallylic halide or the cyclopropyl derivative [15] (see Table 5.4).

Under reaction conditions where the stannane concentration is low the equilibria can establish the most stable secondary acyclic radical prior to

TABLE 5.4

Product yields for the cyclopropylcarbinyl–homoallylic radical rearrangement

Substrate	Stannane concentration	Yield (%)	Products, pent-1-ene: 3-methylbut-1-ene
Cyclopropyl halide	Neat	85	34:66
Cyclopropyl halide	Dilute	85	92:8
Homoallylic halide	Neat	85	10:90
Homoallylic halide	Dilute	85	86:14

hydrogen atom capture and pent-1-ene is formed as the major thermodynamic product. Homoallylic rearrangements of this type, either from the cyclopropyl halide or the open chain system can, accordingly, be predicted to be synthetically useful. Under conditions of kinetic control using neat stannane the homoallylic radical is efficiently quenched to give 3-methylbut-1-ene. More surprisingly, however, treatment of the cyclopropyl halide under similar conditions leads preferentially to rapid ring opening to give the less stable primary radical which is trapped by the stannane. On the reasonable assumption that ring opening is governed by stereoelectronic control, this result would imply that the unique conformation shown is favoured. Further study using a series of derivatives would be most informative.

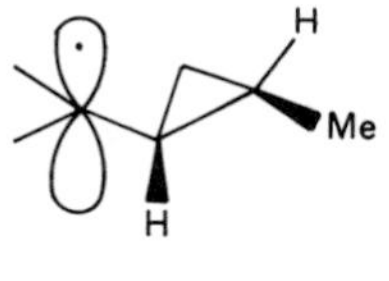

A variety of studies in rigidly constrained bicyclic systems [16] have demonstrated that under kinetic control and irrespective of the alignment of the breaking σ bond, conformational effects then favour a situation leading to highly stereoselective formation of higher energy cycloalkyl methyl radicals, as indicated by the product distributions shown which were not obtained under optimum conditions.

In the case of the bicyclo[4.1.0]heptane derivative shown here, preservation of the norcarane skeleton is possible through the use of triphenyltin hydride as a faster hydrogen atom donor.

More recently, in a detailed study by Clive [17], the use of the phenylseleno group as a low temperature radical trigger for cyclopropyl ring opening of variously substituted bicyclic derivatives was examined. A noteworthy feature of the sequence is the use of a hydroxyl directed Simmons–Smith cyclopropanation to achieve stereospecific creation of a quaternary centre. Examples are given in Table 5.5.

TABLE 5.5

Ring opening reactions of cyclopropylcarbinyl phenylselenides [17]

Substrate (0.1–0.2 M)	Conditions	Product	Yield (%)
	Ph_3SnH, AIBN, benzene, reflux 1 h		92
	Ph_3SnH, sunlamp, Pyrex, toluene, -20 to $-10°C$, 2.5 h		95
	$^nBu_3SnH$, sunlamp, Pyrex, hexane, $0°C$, 2.5 h		86
	$^nBu_3SnH$, sunlamp Pyrex, hexane, $10–30°C$, 2.5 h		65
	$^nBu_3SnH$, AIBN, benzene, reflux		$\geqslant 86$

The addition of arenesulphonyl radicals to vinyl cyclopropanes has also been used by two groups as a convenient entry to 1,5-difunctionalized derivatives.

Thus, readily prepared p-toluenesulphonyl iodide has been used in a thermal reaction for a variety of usefully functionalized 5-iodo-pent-2-enyl sulphones under mild conditions. [18].

The following experimental procedure is typical and additional examples are shown in Table 5.6.

Preparation of 1-(2-iodoethyl)-2-(p-toluenesulphonylmethyl)-
3,4-dihydronaphthalene [18]

TABLE 5.6

Reaction of vinylcyclopropanes with p-toluenesulphonyl
iodide [18].

Substrate	Product	Yield (%)
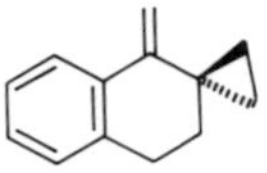	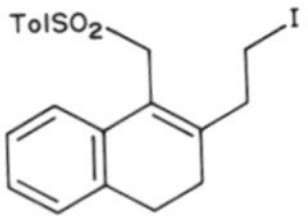 ToISO₂ ... I	95
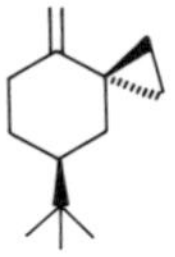	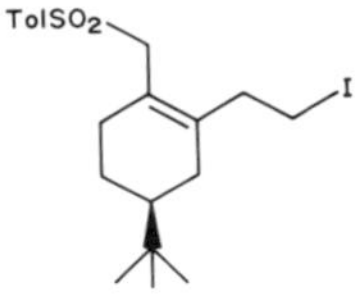 ToISO₂ ... I	59
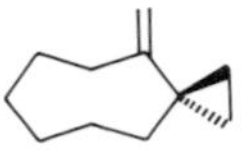	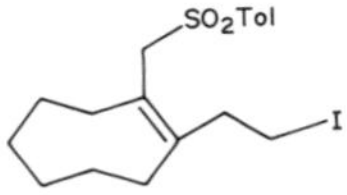 SO₂Tol ... I	52
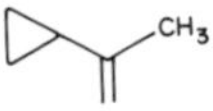	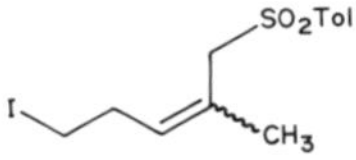 I ... SO₂Tol ... CH₃ (E : Z = 2.4 : 1)	66

A solution of the vinylcyclopropane (135.3 mg, 0.795 mmol) in dichloromethane (1.60 ml) was added over a 30 min period to propylene oxide (0.24 ml)
and p-toluenesulphonyl iodide (246.6 mg, 0.87 mmol) in dichloromethane
(3.6 ml) at 40°C under argon. The mixture was stirred for 2 h, then concentrated *in vacuo* and the residue purified by column chromatography on silica
gel to afford the title product (296.0 mg, 78%) as a white solid, m.p. 33–37°C.

In a similar vein, the congeneric phenylselcno derivative undergoes a photochemically induced reaction, [19].

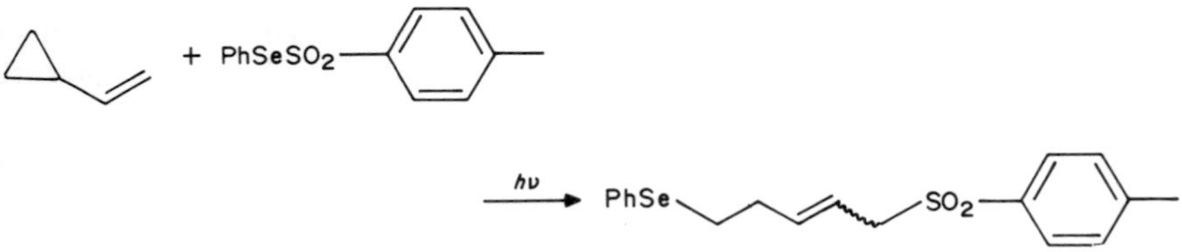

+ PhSeSO₂—⟨aryl⟩—

$\xrightarrow{h\upsilon}$ PhSe⁓⁓SO₂—⟨aryl⟩—

E : Z — 2.4 : 1

The case of cyclopropyl acetylene has also been studied and reveals that capture of the vinylic radical by the reagent competes effectively with cyclopropylcarbinyl radical ring opening [19].

The reaction of vinylcyclopropane itself is typical and additional substrates are collated in Table 5.7.

Selenosulphonation of vinylcyclopropanes—a typical procedure

Se-Phenyl-p-tolueneselenosulphonate (156 mg, 0.50 mmol) and vinylcyclopropane (~ 0.6 mmol) were photolysed in 3 ml of chloroform in a Pyrex vessel for 24 h. Flash chromatography over silica gel (elution with 20% ethyl acetate–hexane) afforded 174 mg (92%) of the 1,5-adduct as a 4:1 mixture of E and Z isomers. Photolyses were performed in a Rayonet RMR-500 reactor equipped with four 254 nm ultraviolet lamps.

Finally, an elegant route to the tropone nucleus involves reduction of suitably substituted dichloro- or dibromomethylcyclohexadienones which are readily prepared from phenols via the Reimer–Tiemann reaction [20].

The sequence is particularly convenient since both the *ortho* and *para* substituted cyclohexadienones give the same tropone.

A mechanism involving ring closure and reopening of the initially formed homoallylic radical generates a carbon centred radical α to the carbonyl group

TABLE 5.7

Photochemical ring opening of vinylcyclopropanes with Se-phenyl-p-tolueneseleno-sulphonate [18]

Substrate	Product	$E:Z$ ratio	Yield (%)
		4:1	92
		8:1	97
		1:10[a]	92
		4:1	95
		5:1	80

[a] Thermal reaction reflux in chloroform with 5 mol% of AIBN for 24 h.

which can then undergo a particularly favourable β-elimination which is driven by the aromaticity of the tropone nucleus.

TABLE 5.8

Preparation of tropones by stannane reduction of dihalogenocyclohexadienones [20]

Phenol	Dihalogenocyclohexadienone	Yield (%)	Tropone or tropolane	Yield (%)
		79		91
		46		25
		37		43

Additional examples are given in Table 5.8 and the communicated experimental procedure [20] consists of "reduction of the cyclohexadienones (1 mmol) with tri-n-butyltin hydride (2.5 mmol) in benzene under reflux for 4 h with catalysis by azobisisobutyronitrile."

Within the cyclopropyl manifold, an isolated example of the generation of a vinyl radical in an alkylidene cyclopropane also indicates that fragmentation to give alkynes is possible [21].

$$^nBu_3SnH$$

Ph—C≡C—

82%

The rapidity of ring opening of the homologous cyclobutylcarbinyl radicals, especially in monocyclic systems, falls off rapidly, and much higher temperatures and low substrate concentrations are required for useful yields [22].

5.5   HOMOBENZYLIC REARRANGEMENTS

A close cousin of the 1,2-vinyl shift induced by the homoallylic–cyclopropylcarbinyl radical series of equilibria is the homobenzylic rearrangement.

The formation of a more stable radical product is, of course, necessary in order to drive the series of ring closure and reopening reactions to completion. Slow capture of the carbon centred radical intermediates by the reagent is also mandatory and dictates how the experiment should be performed.

These considerations may be exemplified in the case of the homobenzylic chlorides shown, where the formation of an intermediate benzyl radical provides a sufficient driving force [23].

42%

The degree of effective experimental control is, however, most clearly exemplified in the case of the homobenzylic halogeno esters shown [24].

		A	B
High stannane concentration	=	77%	23%
Low stannane concentration	=	19%	81%

Rearrangement is particularly favourable in this instance, not only because the rearranged radical is formed α to the ester centre, but also because relief of steric compression can also facilitate the reaction. Nevertheless, by operating at a high stannane concentration, the first formed primary alkyl radical can be effectively trapped to give the kinetic product. Conversely, at low concentrations of tin hydride, sufficient time is given for rearrangement to occur and afford good yields of the thermodynamic product.

The experimental procedures for the above example under kinetic and thermodynamic control are given below, while the related neophyll rearrangement is more properly treated in Chapter 7.

Dehalogenation of methyl 3-bromo-2-methyl-2-phenylpropanoate (kinetic product)

A mixture of tri-n-butyltin hydride (1.6 mmol), the bromide (0.4 mmol) and AIBN (trace) in benzene (160 ml) was refluxed for 4 h under nitrogen. A concentrate of the reaction mixture was passed through a short column of silica gel with chloroform and further purified by preparative TLC (silica gel) using benzene. The products were A and B (A:B = 77:23; 75% yield).

Repetition of the above experiment using tri-n-butylstannane (0.001 mol dm^{-3}) gave A and B in a ratio of 19:81.

From ref. [24] with permission.

5.6 DETERMINENTAL FACTORS IN SOME MISCELLANEOUS RADICAL INDUCED CARBOCYCLIC RING CLEAVAGE REACTIONS

Although relatively unusual, it is important to appreciate that the combination of relief of strain together with the formation of a particularly low energy radical may be sufficient to induce cleavage of a carbon–carbon bond adjacent to a carbon centred radical. The following examples serve to illustrate these considerations.

Thus, fragmentation of the radical derived from the 11-α-xanthate ester of a steroidal cross conjugated ring A dienone led directly to the carbocyclic skeleton of vitamin D derivatives [25].

In this case, the expulsion of a phenoxy radical provides an excellent thermodynamic driving force for reaction to occur.

In similar fashion, 8-bromobornan-2-one has been shown to undergo radical fragmentation to give dihydrocarvone under experimental conditions designed to favour low stannane concentration; at higher concentrations of tin hydride normal dehalogenation is observed [26].

Low stannane concentration	=	92%	8%
High stannane concentration	=	7%	93%

The following experimental details typify the concentrations involved, although from a preparative standpoint slow addition of the stannane and

AIBN to a dilute solution of 8-bromobornan-2-one might be preferable for preparation of dihydrocarvone.

Reaction of 8-bromobornan-2-one with tri-n-butylstannane

To a solution of bromo ketone (200 mg) in benzene (40 ml) was added tri-n-butylstannane (280 mg, 1.1 equiv.) and AIBN (~ 2 mg). The solution was heated at reflux for 24 h, after which time the benzene was removed in a vacuum. The residual oil was dissolved in acetonitrile (30 ml), the solution was thoroughly washed with hexane (5 × 10 ml) to remove any tin compounds and the hexane layer was discarded. Removal of acetonitrile under reduced pressure afforded a yellow oil (0.14 g) which partly crystallized on standing. Gas chromatographic examination (15% FFAP, 180°C) of this oil revealed that it consisted of two components, camphor and dihydrocarvone in a 54:46 ratio. Under more dilute conditions (bromo ketone (100 mg) and tri-n-butylstannane (100 mg) in benzene (250 ml)) the ratio of camphor to dihydrocarvone in the product was found to be 8:92. However, when the same reaction was carried out in concentrated solution (bromo ketone (33 mg) and tri-n-butylstannane (420 mg) in benzene (5 ml)), the predominant product was camphor (93%) while only 7% of the product was dihydrocarvone.

From ref. [26] with permission.

5.7 RING OPENING REACTIONS OF EPOXIDES BY A NEIGHBOURING CARBON CENTRED RADICAL

5.7.1 Introduction

The ring opening of an oxirane by an adjacent carbon centred radical is, as expected by comparison with the cyclopropylcarbinyl analogue (Section 5.4), an extremely rapid process.

However, it differs inasmuch as either carbon–carbon or carbon–oxygen bond cleavage may be observed.

Although synthetically useful reactions have emerged from both types (*vide infra*), the general rule is that in the absence of substituent effects (e.g. vinyl or aryl) which lead to particularly low energy carbon centred radicals, cleavage of the weaker carbon–oxygen bond will prevail. Hence, carbon centred radicals adjacent to an epoxide may be considered as apposite precursors of allylic alkoxyl radicals.

These comments are appropriately demonstrated by early work on the addition of alkyl radicals derived from organoboranes to vinyl oxiranes which established that useful yields of functionalized allylic alcohol derivatives could be formed [27].

Some of the more modern methods which have incorporated this step are discussed below.

5.7.2 Transposition of allylic alcohol derivatives

Radicals derived by stannane reduction of the thiocarbonylimidazolide esters of a variety of alicyclic α,β-epoxy alcohols were shown to undergo efficient carbon–oxygen bond fission to afford rearranged allylic alcohols in a process which represents a formal alternative to the Wharton rearrangement [28], as illustrated for the reaction of the epoxypregnenolone derivative shown here.

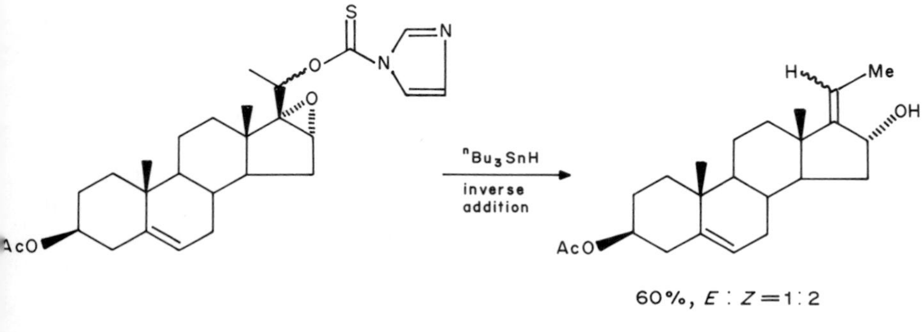

60%, $E : Z = 1 : 2$

Careful experimentation established that the addition of the tri-n-butyltin radical to the thiocarbonyl group always triggered epoxide ring opening and that the most efficient trapping of the resultant allylic alkoxyl radical would therefore result by addition of the substrate to a refluxing solution of the organostannane (inverse addition).

The preparation of the above pregnane derivative is detailed below and additional examples of allylic transposition are collected in Table 5.9.

TABLE 5.9

Allylic transposition by tri-n-butylstannane reduction of α,β-epoxythiocarbonyl-imidazolides [28]

Substrate	Addition mode	Product	Yield (%)
	Normal		65
	Inverse		47
	Inverse		58

Preparation of 3β-acetoxypregna-5,17-dien-16α-ol
(inverse addition method)

A solution of the steroidal thiocarbonyl imidazolide (187 mg, 0.4 mmol) with AIBN (9 mg, 0.06 mmol) in toluene (2 ml) was added dropwise to a solution of tri-n-butylstannane (1.21 g, 4.1 mmol) in toluene (2 ml) at reflux temperature under nitrogen. Reaction was complete when the addition was finished as judged by TLC. The reaction mixture was cooled, carbon tetrachloride (10 ml) was added and stirring was continued until the Sn—H stretch at 1800 cm^{-1} was absent in the infrared spectrum. The mixture was then titrated with a dilute solution of iodine in ether until the iodine colour persisted. After dilution with an equal volume of ether, the reaction mixture was washed with aqueous potassium fluoride until no further precipitation of polymeric tin fluoride was observed. Removal of solvent from the dried organic phase followed by chromatography gave a crystalline mixture of the title alcohols (80 mg, 60%).

From ref. [28] with permission.

In some instances, the capture of the allylic alkoxyl radical proved to be impossible. For example, in the case of the hecogenin derivative shown, the steric inaccessibility of the intermediate 9α-alkoxyl radical and the relief of strain engendered by β-scission led to the ring B opened derivative. Under normal conditions (e.g. low concentrations of stannane) the enone was formed while inverse addition afforded the dihydro derivative by hydrostannylation.

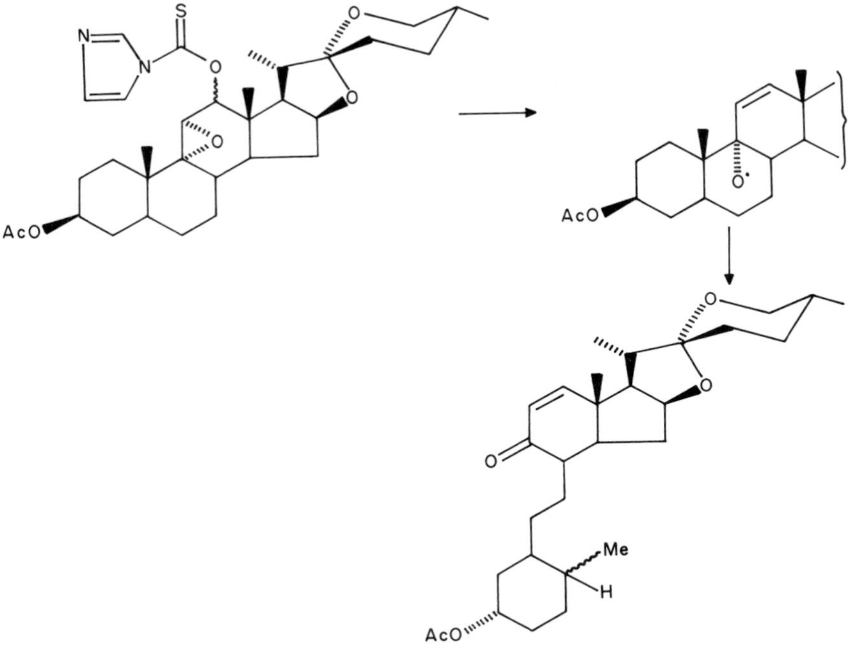

82%

Similarly, by operating at low concentrations of stannane, formation of pinol (18%) was possible through intramolecular addition of the alkoxyl radical to the isopropenyl double bond.

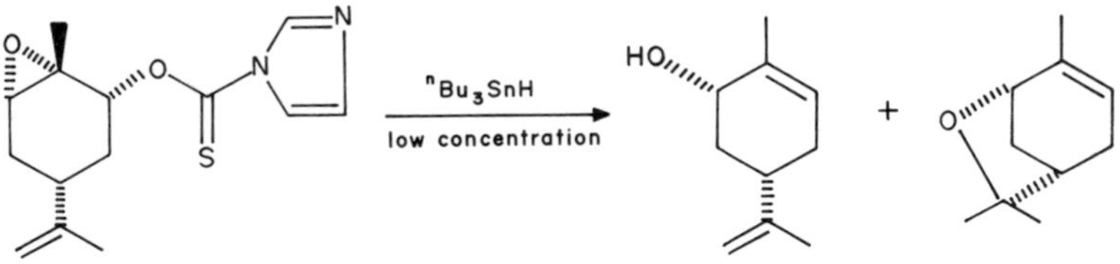

16%   18%

The possibility of controlling the reaction via the addition mode was most convincingly demonstrated in the case of the 4,5-epoxycholestane derivatives. Under conditions of high concentration of stannane the rearranged allylic alcohol is formed by inverse addition. However, addition of stannane to the substrate allowed time for further rearrangement of the alkoxyl via an intermediate or transition state involving β-scission and recombination to give a bridged bicyclic ketone.

A recent addition to this picture has also been provided by the case of the 4α,5α-epoxide which has been shown to rearrange via a reversible β-scission process [29].

The "normal" mode of addition, i.e. operating at low concentrations of stannane, clearly provides a convenient method for the generation of allylic alkoxyl radicals with a view to further study of their rearrangement or addition reactions, prior to hydrogen atom capture.

The preparation of the bridged bicyclic ketone from N-(4β,5-epoxy-5β-cholestan-3α- and 3β-yloxythiocarbonyl)imidazole is typical of the "normal" mode, in terms of concentrations.

Preparation of 5,6-seco-4,6-cyclo-4β-cholestan-5-one (normal addition)

A solution of tri-n-butylstannane (493 mg, 1.7 mmol) and AIBN (14 mg, 0.09 mmol) in dry toluene (9 ml) was added dropwise over 20 min to a refluxing solution of the thiocarbonylimidazolide (427 mg, 0.83 mmol) under nitrogen. The solvent was removed *in vacuo* and the residue chromatographed on silica gel to give the allylic alcohol (78.5 mg, 24%) and the title derivative (150 mg, 47%) as a viscous glass.

5.7.3 Intramolecular cyclization reactions of allyloxy radicals

The intramolecular cyclization of the alkoxy radical to form the strained bicyclic pinol [28] has served as a springboard for a systematic study of intramolecular cyclization reactions of allyloxy radicals to give tetrahydro-furanyl and tetrahydropyranyl derivatives [30].

Cyclization reactions leading from readily available open chain precursors were shown to have promising diastereoselectivity for the formation of 2,2,5-trisubstituted tetrahydrofurans in which the major products featured a *trans* disposition of the two largest alkyl groups.

R	Yield (%)	A	B	C
H		10	4	0
Me		48	8	14
^{n}Bu		56	7	22

The bicyclic ether byproduct C may be formally derived by a tandem addition process which exists in competition with hydrogen atom capture to give A.

For the case of 2,5-disubstituted tetrahydrofurans, however, a uniformly consistent diastereomeric ratio of 3:1 was found in the three cases studied.

Yield (%): R = H(65%), Me (55%), ^{n}Bu (82%)

In the case of tetrahydropyranyl ether formation, a preliminary experiment has provided evidence for high diastereoselectivity, but has also uncovered a competitive intramolecular hydrogen atom abstraction reaction.

A potentially interesting route to medium sized rings has also been discovered by the same group and features intramolecular cyclization of an alkoxyl radical to an alkyne followed by [3,3] sigmatropic rearrangement.

A typical preparation in this series is the cyclization reaction leading to lilac alcohols.

Formation of 3,7-dimethyl-2,3-epoxy-8-t-butyldimethylsilyloxy-6-octenyl-1-oxythiocarbonyl imidazolide and treatment with tri-n-butylstannane and AIBN

The α,β-epoxide (500 mg, 1.7 mmol) and 1,1-thiocarbonyldiimidazole (356 mg, 2.0 mmol) were heated under reflux in dry dichloromethane (30 ml) for 1 h. After cooling, the solvent was removed *in vacuo* to afford the imidazolide as a viscous orange oil. The derivative was dissolved in dry, degassed tetrahydrofuran (50 ml) and heated to reflux with tri-n-butylstannane (1.2 ml, 4.5 mmol) under nitrogen. A solution of AIBN (20 mg) in tetrahydrofuran (2 ml) was added dropwise over 1 h. The mixture was heated under reflux for a further 3 h. On cooling, the solvent was removed *in vacuo* to afford an orange oil. Initial chromatography on alumina with hexane resulted in removal of all imidazole and the majority of the tin residues. A second columning on silica gel with hexane–dichloromethane (1:1) afforded the tetrahydrofuran as a mixture of four diastereoisomers in the form of a colourless oil (217 mg, 45%).

From ref. [30] with permission.

5.7.4 Degradative cleavage of β-vinyl-α,β-epoxycarbonyl compounds

Carbon–oxygen bond cleavage also features in a degradative method for the synthesis of γ-sulphido-α,β-unsaturated aldehydes [31]. In this case the initial carbon centred radical is produced by the classical addition of a thiyl radical to an olefin and the alkoxy radical produced suffers β-scission to give an acyl radical which, in turn, may either undergo decarbonylation (R = t-butyl) or hydrogen atom trapping (R = Ph). Yields of the unsaturated aldehyde are, however, modest.

R,	R¹	Yield (%)
Ph	Me	28
^tBu	Me	27
^tBu	H	19

5.7.5 Reactions involving carbon–carbon bond cleavage in epoxide systems

In remarkable contrast to the behaviour of acyl substituted epoxides (Section 5.7.4), predominant or exclusive carbon–carbon bond fission has been shown to occur in cases where the final carbon centred radical is of particularly low energy [32].

R = phenyl, vinyl

An elegant intramolecular competition experiment using the unsaturated styrene epoxide led to the vinyl ether shown without any detectable formation of cyclopentanoid (or cyclohexane) derivatives. Ring opening to give the benzylic radical is therefore faster than the 5-*exo* cyclization process.

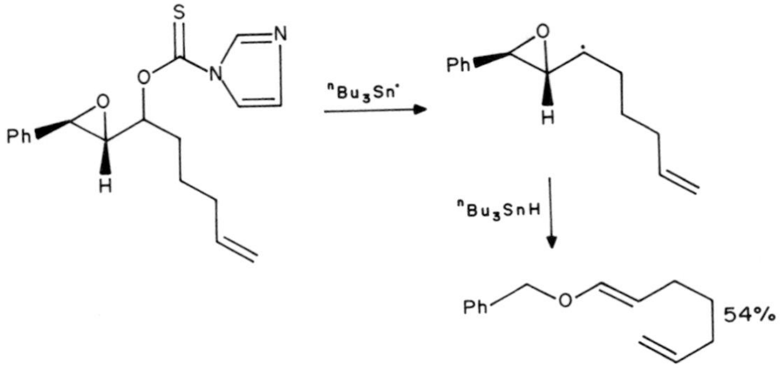

54%

From the preparative standpoint this has led to a novel synthesis of vinyl ethers by treatment of suitably constituted halogeno styrene epoxide precursors.

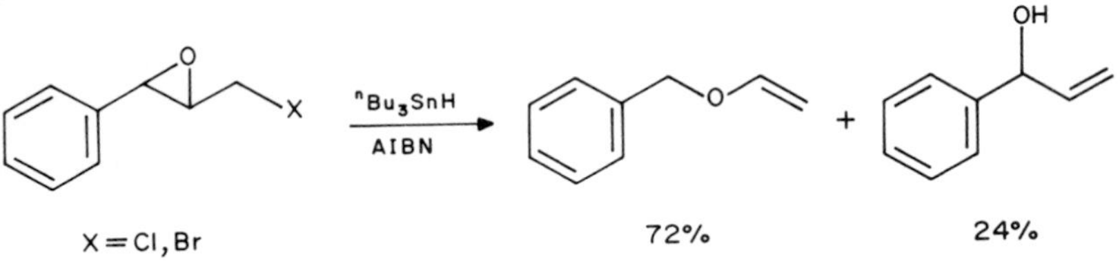

X = Cl, Br   72%   24%

Additional examples studied included the naphthalene diphenyl derivatives shown.

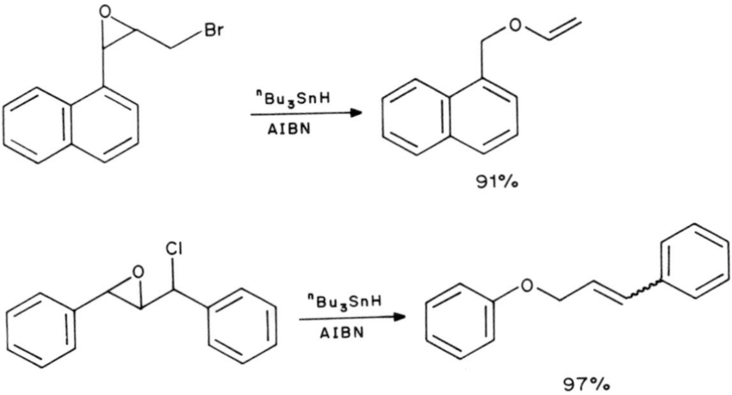

91%

97%

REFERENCES

1. For leading review articles see: A. L. J. Beckwith and K. U. Ingold, in *Rearrangements in Ground and Excited States* (ed. P. de Mayo), Vol. 1, p. 161. Academic Press, New York, 1980; J. W. Wilt, in *Free Radicals* (ed. J. K. Kochi), Vol. II, p. 333. Wiley, New York, 1973.
2. S. Julia and R. Lorne, *C.R. Acad. Sci. Ser. C* **273**, 174 (1971).
3. B. Giese and J. Dupuis, *Tetrahedron Lett.* **25**, 1349 (1984); H.-G. Korth, R. Sustmann, K. S. Groninger, M. Leisung and B. Giese, *J. Org. Chem.* **53**, 4364 (1988).
4. J.-P. Praly, *Tetrahedron Lett.* **24**, 3075 (1983).
5. B. Giese, S. Gilges, K. S. Groninger, C. Lamberth and T. Witzel, *Liebigs Ann. Chem.* 615 (1988); B. Giese, K. S. Groninger, T. Witzel, H.-G. Korth and R. Sustmann, *Angew. Chem. Int. Ed. Engl.* **26**, 233 (1987).
6. C. Chatgilialoglu, D. Griller and M. Lesage, *J. Org. Chem.* **54**, 2492 (1989); M. Lesage, C. Chatgilialoglu and D. Griller, *Tetrahedron Lett.* **30**, 2733 (1989).
7. B. Giese and B. Kopping, *Tetrahedron Lett.* **30**, 681 (1989).
8. H. G. Kuivila, L. W. Menapace and C. R. Warner, *J. Am. Chem. Soc.* **84**, 3584 (1962).
9. M. Julia, C. Rolando and J. N. Verpeaux, *Tetrahedron Lett.* **23**, 4319 (1982).
10. Y. Ueno, H. Sano, and M. Okawara, *Tetrahedron Lett.* **21**, 1767 (1980).
11. For examples see S. J. Cristol, M. W. Klein, M. H. Hendewerk and R. D. Daussin, *J. Org. Chem.* **46**, 4992 (1981); B. E. Cross, A. Erasmuson and P. Filippone, *J. Chem. Soc. Perkin Trans. 1*, 1293 (1981).
12. T. Cohen and M.-T. Lin *J. Am. Chem. Soc.* **106**, 1130 (1984).
13. Y. Ueno, S. Aoki and M. Okawara, *J. Am. Chem. Soc.* **101**, 5414 (1979).
14. K. U. Ingold, *Pure Appl. Chem.* **56**, 1767 (1984).
15. M. Castaing, M. Pereyre, M. Ratier, P. M. Blum and A. G. Davies, *J. Chem. Soc. Perkin Trans. II* 287 (1979); M. Ratier, M. Pereyre, A. G. Davies and R. Sutcliffe, *J. Chem. Soc. Perkin Trans. II* 1907 (1984); P. M. Blum, A. G. Davies, M. Pereyre and M. Ratier, *J. Chem. Soc. Chem. Commun.* 814 (1976); A. Effio, D. Griller, K. U. Ingold, A. L. J. Beckwith and A. K. Serelis, *J. Am. Chem. Soc.* **102**, 1734 (1980).
16. S. J. Cristol and R. V. Barbour, *J. Am. Chem. Soc.* **90**, 2832 (1968); E. C. Friedrich and R. L. Holmstead, *J. Org. Chem.* **37**, 2546, 2550 (1972); A. L. J. Beckwith and G. Moad, *J. Chem. Soc. Perkin Trans. II* 1473 (1980).
17. D. L. J. Clive and S. Daigneault, *J. Chem. Soc. Chem. Commun.* 332 (1989).
18. A. D. Morris, M. C. de C. Alpoim, W. B. Motherwell and D. M. O'Shea, *Tetrahedron Lett.* **29**, 4173 (1988).
19. T. G. Back and K. R. Muralidharan, *J. Org. Chem.* **54**, 121 (1989).
20. M. Barbier, D. H. R. Barton, M. Devys and R. S. Topgi, *J. Chem. Soc. Chem. Commun.* 743 (1984).
21. A. Rahm and M. Degueil-Castaing, Unpublished results cited in *Tin in Organic Synthesis* (ed. M. Pereyre, J.-P. Quintard and A. Rahm), p. 64, ref. 188. Butterworths, London, 1987.
22. M. Castaing, M. Pereyre, M. Ratier, P. M. Blum and A. G. Davies, *J. Chem. Soc. Perkin Trans. II* 287 (1979); A. L. J. Beckwith and G. Moad, *J. Chem. Soc. Perkin Trans. II* 1083 (1980).
23. B. B. Jarvis and J. B. Yount, *J. Org. Chem.* **35**, 2088 (1970).
24. M. Tada, S. Akinaga and M. Okabe, *Bull. Chem. Soc. Jpn.* **55**, 3939 (1982).
25. T. P. Ananthanarayan, T. Gallaher and P. D. Magnus, *J. Chem. Soc. Chem. Commun.* 709 (1982); T. P. Ananthanarayan, P. D. Magnus and A. W. Norman, *J. Chem. Soc. Chem. Commun.* 1096 (1983).
26. D. P. G. Hamon and K. R. Richards, *Aus. J. Chem.* **36**, 109 (1983).

27. H. C. Brown and M. M. Midland, *J. Am. Chem. Soc.* **93**, 4078 (1971).
28. D. H. R. Barton, R. S. Hay-Motherwell and W. B. Motherwell, *J. Chem. Soc. Perkin Trans. 1* 2363 (1981).
29. W. R. Bowman, B. A. Marples and N. A. Zaidi, *Tetrahedron Lett.* **30**, 3343 (1989).
30. A. Johns and J. A. Murphy, *Tetrahedron Lett.* **29**, 837 (1988); A. Johns, J. A. Murphy and M. S. Sherburn, *Tetrahedron* **45**, 7835 (1989).
31. J. A. Murphy, C. W. Patterson and N. F. Wooster, *Tetrahedron Lett.* **29**, 955 (1988).
32. M. Cook, O. Hares, A. Johns, J. A. Murphy and C. W. Patterson, *J. Chem. Soc. Chem. Commun.* 1419 (1986); A. Johns, J. A. Murphy, C. W. Patterson and N. F. Wooster, *J. Chem. Soc. Chem. Commun.* 1238 (1987).

REFERENCES

−6−

Intermolecular Carbon–Carbon Bond Forming Free Radical Chain Reactions

6.1 ADDITION OF CARBON CENTRED RADICALS TO MULTIPLE BONDS

Carbon centred radicals undergo addition to alkenes via an unsymmetrical transition state which occurs at an early stage on the reaction coordinate. Calculations for the addition of a methyl radical to ethene suggest that the incoming radical approaches along a trajectory perpendicular to the nodal plane of the π system at approximately the tetrahedral angle and that the separation in the transition state is around 2.3 Å [1].

$$2.3\,\text{Å} \quad 109° \quad C_\alpha \quad C_\beta$$

Related calculations for the addition of a methyl radical to ethyne predict a similar separation at the transition state and an approach angle of 114° [1].

$$2.3\,\text{Å} \quad 114° \quad H—C_\alpha \equiv C_\beta—H$$

Accordingly, radical additions to alkenes and alkynes are subject to frontier molecular orbital control with nucleophilic radicals adding preferentially to electron deficient alkenes and electrophilic radicals to electron rich alkenes (Section 1.3) [2].

A further consequence of the unsymmetrical nature of the transition state is the relatively high susceptibility of the reactions to steric hindrance at the α- but not the β-position of the alkene or alkyne (Table 6.1).

Effectively, and for all practical purposes, nucleophilic radicals only add efficiently to terminal electron deficient alkenes. Addition to internal alkenes is possible when the substituent at the α-position is a second electron withdrawing group. Thus, nucleophilic radical addition to fumarate esters

TABLE 6.1

Comparison of the effects of α- and β-substituents on the rate of
addition of the cyclohexyl radical to alkenes at 20°C

Z or Y	k_β	k_α	k_β/k_α
CN	310	6.0	51
CO_2Me	150	5.0	30
Cl	10	0.067	149
Ph	6.4	0.009	710
H	1.0	1.0	1.0
Me	0.71	0.011	64
Et	0.55	0.007	83
^{i}Pr	0.43	0.002	290
^{t}Bu	0.24	5.0×10^{-5}	4800

Adapted from B. Giese, *Angew. Chem. Int. Ed. Engl.* **22**, 753 (1983).

for example, and especially to cyclic doubly activated alkenes such as maleic anhydride and maleimides, is a synthetically viable process.

The early nature of the transition state also dictates that, for nucleophilic radical addition to electron deficient alkenes, the rate of addition is related to the electron withdrawing capacity of the β-substituent and not to any ability to stabilize the adduct radical.

In the attacking radical, branching at the α-position (the radical centre) increases the nucleophilicity of the radical by the inductive effect and so increases the rate of attack at a given electron deficient alkene. For example, the t-butyl radical is 24 times more reactive than the methyl radical towards diethyl vinylphosphonate at $-40°C$ [3]. Branching at the β-position of the attacking radical, however, leads to increased susceptibility to steric hindrance.

6.2   STEREOCHEMICAL ASPECTS [4]

Simple alkyl radicals are planar, sp^2 hybridized species and, in the absence of further stereogenic centres, reaction is equally likely at either face. The ability of further stereogenic centres in a radical to induce diastereofacial selectivity in the reactions of the radical is restricted in acyclic systems by free rotation about carbon–carbon bonds. This restriction applies equally to both the attacking and the adduct radicals. The obtention of synthetically useful diastereoselectivities in acyclic systems is only possible by the limitation of conformational mobility, as demonstrated by Porter with an amide linkage [5].

Diastereoselectivity in the reactions of unsymmetrical alicyclic radicals is a common phenomenon. Thus, addition of the 2-ethoxycyclopentyl radical to acrylonitrile by the mercury method gives predominantly the *trans* disubstituted cyclopentane [6].

In the corresponding cyclohexyl radical, lower selectivity (70:30) is observed. Particularly high diastereoselectivities are obtained when the radical forms part of a [3.3.0]bicyclooctane system [7].

Cyclic [8] and bicyclic [9] systems may be temporarily set up by *inter alia* the formation of isopropylidene derivatives of vicinal diols.

The principle of using ring systems to induce selectivity is readily extended to the radical trap as illustrated by the studies of Lefort on the addition of bromotrichloromethane to esters of fumaric and maleic acids and to maleic anhydride [10]. Both diethyl maleate and fumarate gave the same 3:1 ratio of *erythro:threo* adducts but the *threo* adduct was the exclusive product with maleic anhydride.

However, a cautionary note must be sounded here: diastereoselectivity in radical addition to substituted maleic anhydrides is dependent on the bulk of the attacking radical [11].

R=Me	43	:	57
R=^{t}Bu	92	:	8

6.3 ADDITION TO ALKENES WITH CHAIN TRANSFER BY HYDROGEN ATOM ABSTRACTION

6.3.1 Direct addition of activated carbon–hydrogen bonds

The overall addition of an alkyl radical R˙ and a hydrogen atom H˙ across a double bond can be achieved by simply heating or photolysing an alkene and an excess of the appropriate substrate R—H in the presence of a free radical initiator. A general chain mechanism can be written for this process.

The method is limited by the high strength of typical carbon–hydrogen bonds which slow down the chain transfer step to such an extent that large excesses of R—H have to be used if alkene polymerization is to be avoided. Only those substrates containing relatively activated C—H bonds are useful and even then only when so readily available as to permit their use as the reaction solvent. Table 6.2 summarizes some useful substrates, their sites of attack and the nature of the products formed.

TABLE 6.2

Substrates for direct radical addition to alkenes

Substrate	Products
	Aldehydes
	Esters
	Amides
	Ketones
	Lactams (a:b ≃ 2:1)
	Lactones (a:b ≃ 1:20)
	Ethers
	Amines
Cl_3C—H ⟵	Trichloroalkanes

In certain cases where the substrate is readily available such methods can constitute simple effective syntheses.

Such direct additions of C—H bonds across alkenes have been extensively studied and reviewed [12] and form the subject of an article in the *Organic Reactions* series [13].

6.3.2 The mercury method

In the mercury method an alkylmercury hydride is generated *in situ* by the reduction of alkylmercury halides or acetates with a borohydride reducing agent.

$$R—Hg—X + NaBH_4 \rightarrow R—Hg—H$$

$$X = Cl, Br, OAc$$

The alkylmercury hydride acts as a source of both the alkyl radical and the hydrogen atom. A simple chain mechanism may be written. The reactions are autoinitiating.

Both the carbon–mercury and the mercury–hydrogen bonds are weak and the reactions proceed readily at room temperature.

General procedure for the reaction of alkylmercury acetates (or halides) with acrylonitrile and sodium borohydride [14]

The alkylmercury acetate (3 mmol) and acrylonitrile or methyl acrylate (3–10 mmol) were dissolved in ethanol (20 ml) and treated at 0°C under a nitrogen atmosphere with sodium borohydride (3 mmol) in ethanol or water (5 ml). When precipitation of mercury was complete the reaction mixture was filtered on celite, the solvents removed *in vacuo* and the residue taken up in chloroform and washed with 2 M sodium hydroxide. After drying and removal of the solvent the adduct was isolated by kugelrohr distillation or chromatography on silica gel.

The greatest attribute of this method is the wide variety of methods available for the generation of organomercury halides and acetates [15]. Thus, in the simplest case an alkyl halide is converted via the corresponding Grignard reagent into the alkylmercury halide which is then coupled by the standard procedure. The method is compatible with most functional groups except, of course, those susceptible to rapid reduction by borohydrides. Complex functionality may be accommodated in the alkene as well as in the organomercury halide or acetate provided that the alkene is sufficiently electron deficient, as illustrated by the application of the method to small peptides [16].

The well known procedures [15] of alkoxy- and acetoxymercuration of double bonds permit facile entry into β-oxygenated alkylmercury derivatives which are then readily coupled with electron deficient alkenes as illustrated for triacetylglucal [17].

60%, α:β = 2:1

This sequence is applicable to any alkene capable of undergoing acetoxy or alkoxymercuration and may be carried out without isolation of the intermediate acetoxy- or alkoxymercurial.

Typical procedure for the methoxymercuration/borohydride coupling to acrylonitrile [18]

Mercuric acetate (4.14 g; 13 mmol) and cyclopentene (2.04 g; 30 mmol) in methanol (10 ml) were treated with mercuric oxide (1.52 g; 7.0 mmol) in four portions at 20°C. When the solution turned orange/red, dichloromethane (100 ml), 2 M sodium hydroxide (1 ml) and acrylonitrile (3.18 g, 60 mmol) were added. After cooling to 0°C, dichloromethane (100 ml) and sodium borohydride (1.53 g, 40 mmol) were added and the reaction stirred for 1 h. The excess borohydride was then dissolved by the addition of water (10 ml) and the aqueous phase further extracted with dichloromethane (2 × 20 ml). The combined organic phases were dried on magnesium sulphate. Removal of the solvent and distillation yielded a 22:78 *cis/trans* mixture of 3-(2-methoxycyclopentyl)propiononitrile in 65% yield.

Use of an appropriately substituted alkene permits formation of a cyclic ether or lactone in the oxymercuration step which can then be coupled to an electron deficient alkene in the usual way [19].

X=H₂, 44%
X=O, 40%

A further extension of this methodology involves the coupling of a β-acetoxyalkylmercury acetate with acrylonitrile followed either by treatment with sulphuric acid giving a γ-lactone or by treatment with sodium hydroxide giving a δ-lactone [20].

Amino-, amido- and ureidomercuration of alkenes leading to β-nitrogen substituted alkylmercury derivatives is also a useful sequence when used in conjuction with the borohydride mediated addition to electron deficient alkenes [21, 22].

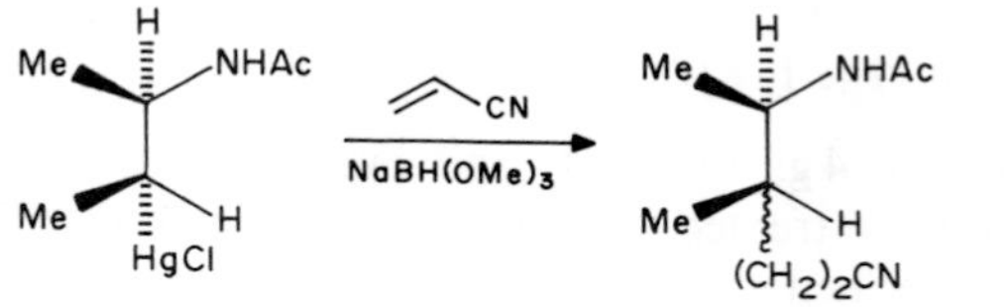

74 %, *erythro* : *threo* = 32 : 68

Unfortunately, the procedure is not applicable to β-azido- or β-nitro-alkylmercury derivatives due to the rapid β-elimination of the azido and nitrogen dioxide radicals, respectively, from the intermediate radical [21]. The aminomercurial can take the form of a cyclic system generated by intramolecular aminomercuration [19, 21, 23]. It should, however, be noted that the method is not compatible with butyrolactams owing to ring opening [21].

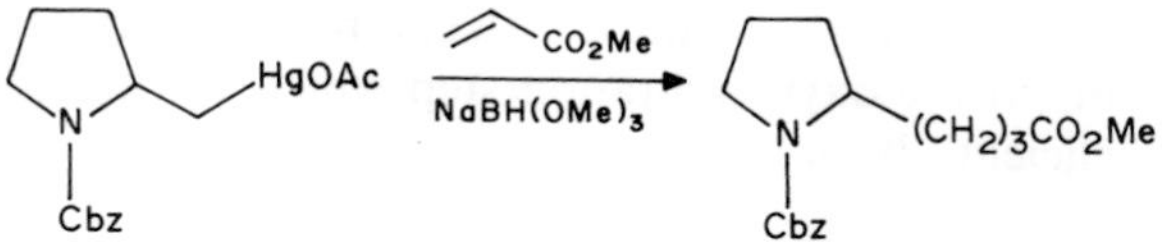

64%

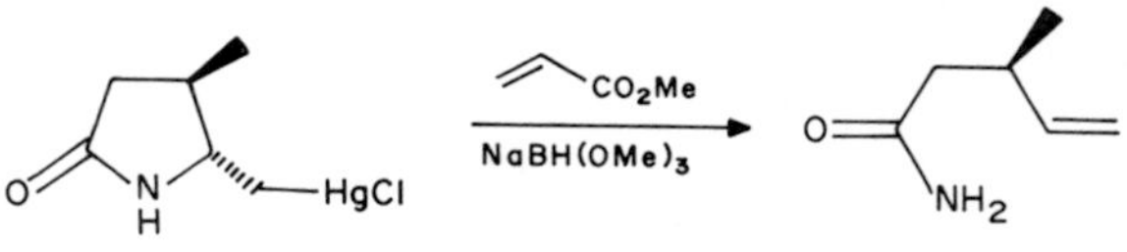

100%

As with oxymercuration, the intermediate amino- or amidomercurials can, but need not be, isolated; high yielding examples of their generation and *in situ* coupling are described in the literature [23].

γ-Alkoxy radicals for addition to activated alkenes are generated by oxymercuration of cyclopropanes followed by reduction with borohydride in the presence of the alkene. Ring opening of cyclopropanes in this manner is regiospecific with attack by mercuric acetate at the least substituted ring position and ring opening towards the most substituted position.

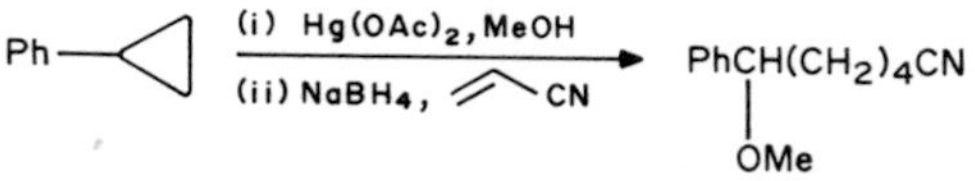

87%

Once again it is possible to isolate the alkoxymercurial or to execute the complete mercuration/coupling sequence in one pot [24].

An interesting variation on the cyclopropane theme is the use of trimethylsilyloxycyclopropanes leading after oxymercuration to 3-oxomer-curials and hence to 3-oxo radicals. In a typical sequence [25] an aldehyde or ketone is converted to its trimethylsilyl enolether which is then subjected to the Simmons–Smith cyclopropanation reaction. Exposure to mercuric acetate yields the 3-oxomercurial which is coupled to an electron deficient alkene in the usual manner.

Electron rich alkenes may be coupled reductively at their least substituted position to electron deficient alkenes by a hydroboration/transmetallation procedure [26].

This procedure which consumes 3 mol of alkene for the generation of 1 mol of organomercurial is evidently wasteful of the alkene. A more efficient procedure relies on the selective transmetallation of primary alkyl–boron bonds and employs dicyclohexylborane in the hydroboration step [26].

Coupling of electron poor and electron rich alkenes by
hydroboration/mercuration

A solution of cyclohexene (1.81 g, 22 mmol) in tetrahydrofuran (5 ml) was
added over 5–20 min to a 1 M solution of borane in tetrahydrofuran (11 ml,
11 mmol) under nitrogen at 0°C. After 2 h the electron rich alkene (10 mmol)
was added and the reaction stirred for 12 h at 20°C. The reaction mixture
was then treated with mercuric acetate (3.18 g, 10 mmol) and, after stirring
for 30 min, with the electron poor alkene (20–50 mmol). Sodium borohydride
(450 mg, 12 mmol) suspended in water (1 ml) was then added leading to the
precipitation of mercury within 5 min. Organoborane residues were removed
by silica gel chromatography or oxidation with hydrogen peroxide and
aqueous work up. Pure coupled products were then isolated by distillation.

From ref. [26] with permission.

Finally in this section it is appropriate to outline the limitations of the
method. Alkylmercury hydrides are highly efficient hydrogen atom donors
and, as such, attempted use of non-terminal alkenes, other than those
activated at both termini, simply results in reductive demercuration of the
organomercurial. In principle, any organomercurial may be used including
bridgehead and cyclopropylmercurials, leading to σ-carbon radicals [27, 28].
However, attempted use of arylmercurials is reported to be unsuccessful [28].

6.3.3 The tin hydride method

In the tin hydride method an alkyl iodide or pseudohalide is reacted with a
trialkyl- or triaryltin hydride in the presence of an alkene and AIBN as the
initiator. The method was first described by Burke and coworkers [7] who
studied the tri-n-butyltin hydride mediated coupling of alkyl iodides and
alkyl phenylselenides with methyl acrylate and methyl methacrylate. In all
cases studied, better yields were obtained from the phenylselenides than
from the iodides. This is not, however, necessarily indicative of a cleaner
radical reaction but may be related to the greater ease of purification found
with the selenides (Section 1.7). Interestingly, both the alkyl iodides and
phenylselenides were observed to give significantly higher yields than when
the identical coupling was carried out by the mercury method [7].

X = I, 54%   X = SePh, 70%

*Tri-n-butyltin hydride mediated coupling of alkyl halides and
pseudohalides with electron deficient alkenes* [29]

Tri-n-butyltin hydride (0.8 mmol) and a crystal of AIBN in toluene (1.0 ml)
were added via a syringe pump over 10 h to a solution of the alkyl iodide
or phenylselenide (0.4 mmol) and methyl acrylate (2 mmol) in toluene at 100°C
under a nitrogen atmosphere. The solvent was removed *in vacuo* and the
product isolated by chromatography on silica gel. In many instances the
removal of organostannane residues prior to chromatography could be
advantageously achieved with one or other of the methods outlined in Section
1.7, although this was not usually necessary with the selenides.

The tin hydride method which proceeds via a simple chain mechanism
may in principle by applied to a variety of pseudohalides as described for

$$R \cdot \; + \; \overset{}{\diagup}\!\!\!\diagdown_{Z} \; \longrightarrow \; R \diagdown\!\!\!\diagup\!\!\!\diagdown_{Z} \cdot$$

$$R \diagdown\!\!\!\diagup\!\!\!\diagdown_{Z} \cdot \; + \; H\!\!-\!\!Sn^{n}Bu_3 \; \longrightarrow \; R \diagdown\!\!\!\diagup\!\!\!\diagdown_{Z} \; + \; {}^{n}Bu_3Sn \cdot$$

$${}^{n}Bu_3Sn \cdot \; + \; R\!\!-\!\!X \; \longrightarrow \; R \cdot \; + \; {}^{n}Bu_3SnX$$

the removal of functional groups in Chapter 3. However, in practice only
those compounds R—X in which the abstraction of X by the stannyl radical
is sufficiently rapid to compete with hydrostannylation of the alkene are
useful. Thus the reaction has been applied to alkyl bromides and iodides, to
alkyl phenylselenides, to thionocarbonyl esters and to tertiary nitroalkanes
with success. Table 6.3 gives some representative examples chosen so as to
illustrate both the variety of useful radical leaving groups and the structural
complexity and diverse functionality compatible with the method.

An interesting modification of this method is the incorporation of the
radical accepting alkene into the substrate molecule with the use of a simple
alkyl halide as the radical precursor. The yield obtained is remarkable given
the internal nature of the alkene [36].

64%

TABLE 6.3

The tin hydride method for the addition of alkyl radicals to alkenes

Substrate	Alkene	Products	Yield (%)	Ref.
(acetylated pyranose–SePh)	$CH_2=CHCO_2Me$	(pyranose–(CH$_2$)$_2$CO$_2$Me)	40	[30]
(acetylated pyranose–Br)	$CH_2=CHCN$	(pyranose–(CH$_2$)$_2$CN)	65	[31]
(PhCH$_2$O$_2$C, HNCbz, H)—I	$CH_2=CHCO_2H$	PhCH$_2$O$_2$C(HNCbz, H)—(CH$_2$)$_3$CO$_2$H	30	[30]
(BzO pyranose–I, OMe)	(HO, OH dilactone)	(lactone–CH$_2$–pyranose OMe, OBz)	35[a]	[32]
(Br, OAc acetylated sugar)	$CH_2=CHCN$	(38) + (pyranose products)	35	[33]
(acetonide thiocarbonate SMe)	$CH_2=CHCN$	(furanose–(CH$_2$)$_2$CN) β:α = 3:1	40	[34]
(Ph, NO$_2$, CN nitro compound)	$CH_2=C(CO_2Me)$	(Ph, CO$_2$Me, CN product)	46	[35]

[a] After acetylation of the crude reaction mixture.

A recent development is the use of acyl phenylselenides in conjunction with tri-n-butyltin hydride as acyl radical precursors [37]. This reaction, which is applicable to primary alkyl and arylacyl phenylselenides, but not to secondary or tertiary alkylacyl phenylselenides owing to competing decarbonylation, is a practical alternative to the addition of aldehydes to alkenes (Section 6.3.1).

The requisite acyl phenylselenides are readily obtained by the reaction of acyl chlorides with sodium phenylselenide or by simply stirring the acid with tri-n-butylphosphine and phenylselenenyl bromide [38].

As with the mercury method the tin hydride method suffers from one major drawback, namely the efficiency of the tin hydride as a hydrogen atom donor. A serious competing pathway therefore is simple reduction of the radical precursor. In order to overcome this problem it is usual to add the tin hydride slowly into the reaction mixture containing an excess of the alkene. Evidently this is a recipe for alkene telomerization and in practice a compromise between reduction, tin hydride concentration, telomerization and product formation has to be found.

In view of this problem and the related problem of the removal of organotin residues from the crude product (Section 1.7), substantial effort has been devoted to the development of alternatives to tin hydrides. The simplest procedure consists of the use of a catalytic quantity of tri-n-butyltin hydride or chloride in the presence of a borohydride reagent as the overall reductant. In this manner the tin hydride concentration is maintained at a minimum and the purification problem solved.

Several examples of this process have been recorded [34]. Yields for primary secondary and tertiary alkyl iodides with terminal activated alkenes are good.

An alternative procedure, developed for preparative chemistry by Pike and coworkers [39], replaces tri-n-butyltin hydride by tri-n-butylgermanium hydride, a less efficient hydrogen atom donor.

$$R—I + \overset{}{\underset{Z}{\diagup\!\!\diagup}} \xrightarrow[80°C]{^nBu_3GeH, AIBN} R\diagdown\!\!\diagup\!\!\diagdown_Z$$

A limitation of this reagent, however, is the reduced rate of halogen abstraction from alkyl halides by the tri-n-butylgermyl radical which can result in poor chain propagation. A further problem is the prohibitive cost of trialkylgermanium derivatives.

A further alternative, having similar hydrogen donor and chain propagating properties to tri-n-butyltin hydride, is tris(trimethylsilyl)silane. This reagent which can also be used catalytically in conjuction with sodium borohydride was designed and is reported to have similar properties to tri-n-butyltin hydride but to side step the troublesome purification [40].

6.3.4 The trialkylborane method

Hydroboration of an alkene, followed by admixture with an $\alpha\beta$-unsaturated carbonyl compound and exposure of the mixture to traces of oxygen or other free radical initiator, results in the overall coupling of the alkene with the $\alpha\beta$-unsaturated carbonyl compound [41].

$$3RCH{=}CH_2 \xrightarrow{B_2H_2} (RCH_2CH_2)_3B$$

$$(RCH_2CH_2)_3B + CH_2{=}CHCOX \xrightarrow[\text{(ii) } H_3O^+]{\text{(i) init.}} R(CH_2)_4COX$$

This reaction, which proceeds via a simple chain mechanism, differs fundamentally from the mercury and tin methods in so far as the chain transfer step does not involve hydrogen abstraction by the adduct radical from a suitable hydrogen atom donor. Chain transfer is accomplished by S_H2 at the boron centre and the product is a dialkylborinyl enolate.

The carbonyl compound is generated on hydrolysis. The absence of an *effective* competing chain sequence enables the use of internal alkenes as radical traps [42].

(i) BH₃

(ii)

(iii) H₂O

Coupling of cyclohexene with 3-penten-2-one by hydroboration [43]

A 25 ml flask, fitted with a nitrogen inlet, septum cap and condenser was flushed with nitrogen and charged with borane (11 mmol) in tetrahydrofuran (10 ml). Cyclohexene (3.3 ml, 33 mmol) was added and the mixture stirred at 50°C for 3h to ensure complete formation of tricyclohexylborane. Water (0.36ml, 20 mmol) followed by 3-penten-2-one (1.42, 10 mmol) was added and air passed into the flask at the rate of 1 ml min^{-1} through a syringe needle to a point just above the surface of the reaction mixture. The reaction was stirred at room temperature and monitored by gas chromatography. After completion the reaction was worked up by treatment with hydrogen peroxide and extraction.

A more efficient procedure, in terms of alkene converted to product, involves the use of ethoxydiphenylborane in the hydroboration step [44].

$$RCH{=}CH_2 + Ph_2BOEt \xrightarrow{\text{LiAlH}_4} RCH_2CH_2BPh_2$$

$$RCH_2CH_2BPh_2 + CH_2{=}CHCOMe \xrightarrow[\text{(ii)}\,H_2O]{\text{(i)}\,O_2} R(CH_2)_4COMe$$

6.4 ADDITION TO ALKENES WITH CHAIN TRANSFER BY FORMATION OF A CARBON–HETEROATOM BOND

6.4.1 The *O*-acyl thiohydroxamate method

Photolytically or thermally initiated decomposition of *O*-acyl thiohydroxa-mates in the presence of terminal electron deficient alkenes results in the establishment of a chain reaction and formation of the adducts of R˙ and ˙S–2-py across the alkene.

$$R\text{—}CO_2\cdot \longrightarrow R\cdot + CO_2$$

Such addition reactions are simple to carry out and proceed at or below room temperature on white light photolysis or simply in benzene or toluene at reflux [45]. Decarboxylative rearrangement (Section 3.7.3) is, however, a competing background reaction which precludes the use of all but terminal activated alkenes as radical traps (with the exception of doubly activated alkenes and nitroalkenes). Table 6.4 collects together some examples which illustrate the compatibility of diverse functional groups and a variety of different electron deficient alkenes.

The doubly functionalized carbon centres generated in these reactions make them particularly attractive from a synthetic point of view. Thus the pyridylthio moiety may simply be removed reductively with Raney nickel or tri-n-butyltin hydride. Alternatively, controlled oxidation affords sulphoxides which may take place in *syn* elimination reactions on heating.

TABLE 6.4

The O-acyl thiohydroxamate method for radical addition to alkenes

Substrate	Alkene	Product (Yield %)	Ref.
$Me(CH_2)_{14}CO_2H$	CO₂Me	$Me(CH_2)_{15}CH(CO_2Me)S$-py (83)	[45]
$Me(CH_2)_{14}CO_2H$	CN	$Me(CH_2)_{15}CH(CN)S$-py (57)	[45]
		(100)	[46]
$(PhCH_2)_2CHCO_2H$	SO₂Ph	$(PhCH_2)_2CHCH_2CH_2(SO_2Ph)S$-py (75)	[47]
$CH_3(CH_2)_{14}CO_2H$		$CH_3(CH_2)_{14}CH(CN)CH(CN)S$-py (60)	[45]
$CH_3(CH_2)_{14}CO_2H$		(30)	[45]
		(78)	[46]
	CO₂Me	(62)	[48]

O-Acyl thiohydroxamate method for radical addition to activated alkenes [49]

Photochemical. The acid chloride (1 mmol) in benzene (1 ml) was added to a stirred solution of 2-mercaptopyridine-N-oxide (140 mg, 1.1 mmol) and pyridine (0.25 ml) in benzene (10 ml) at room temperature under nitrogen. The reaction mixture was stirred for 30 min and then filtered to remove the precipitated pyridinium hydrochloride. The olefin (3–5 mmol) was then added and the solution irradiated (300 W tungsten lamp) at room temperature (water bath) under nitrogen with stirring. After completion (TLC; disappearance of yellow coloration) the solvent was removed under reduced pressure and the products isolated by chromatography on silica gel.

Thermal. A solution of the *O*-acyl thiohydroxamate (1 mmol), either prepared *in situ* as above or pure isolated [46], was added over 5 min to a solution of the alkene in toluene (10 ml) at reflux under nitrogen. After completion the products were isolated by chromatography on silica gel.

Use of nitroalkenes as radical traps in conjunction with the *O*-acyl thiohydroxamates leads to the formation of α-nitrosulphides which may be converted oxidatively [45, 46] into carboxylic acids, so providing a radical equivalent to the Arndt–Eistert reaction, or "hydrolysed" to aldehydes [46].

Addition to vinylsulphones [47] provides α-(pyridylthio)sulphones which may be transformed into a variety of functional groups.

Addition to cyclic doubly activated alkenes such as maleic anhydride is efficient. However, under the reaction conditions elimination of the heterocyclic sulphide occurs and so provides an extremely simple entry into

substituted maleic anhydrides [45]. A similar reaction is observed with
p-quinones [45, 50].

Protonated heteroaromatic bases also serve as radical traps for radicals
derived from O-acyl thiohydroxamates yielding, ultimately, alkylated
heteroaromatic bases. It is suggested that the primary reaction product
undergoes elimination of pyridin-2-thione to give the observed products. The
reactions are carried out by white light photolysis of the O-acyl
thiohydroxamate in dichloromethane in the presence of the base as its
camphor-10-sulphonate salt [51].

O-Acyl thiohydroxamates may also be prepared from perfluoroalkanoic
acids, enabling the addition of perfluoroalkyl radicals to electron rich alkenes
without recourse to the corrosive perfluoroalkyl iodides [52].

Finally, the O-acyl thiohydroxamate methodology for the generation of tertiary radicals by double decarboxylation of half-oxalate esters (Chapter 3) is also applicable to the formation of carbon–carbon bonds [53].

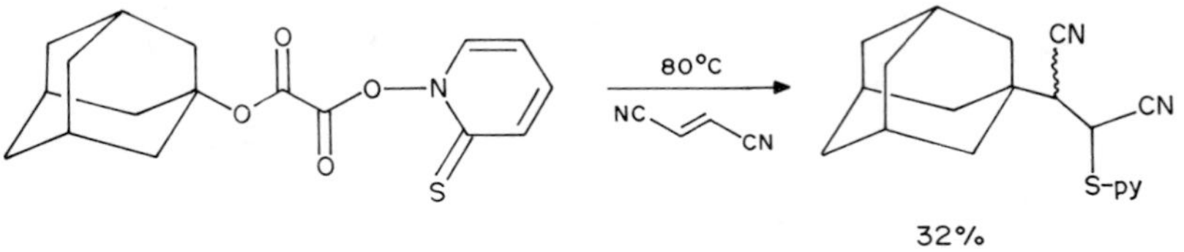

32%

6.4.2   The halogen atom transfer method

The addition of perhaloalkanes to alkenes by a radical chain mechanism in which chain transfer is achieved by halogen atom abstraction from the substrate has been known for many years and has been reviewed in *Organic Reactions* [13].

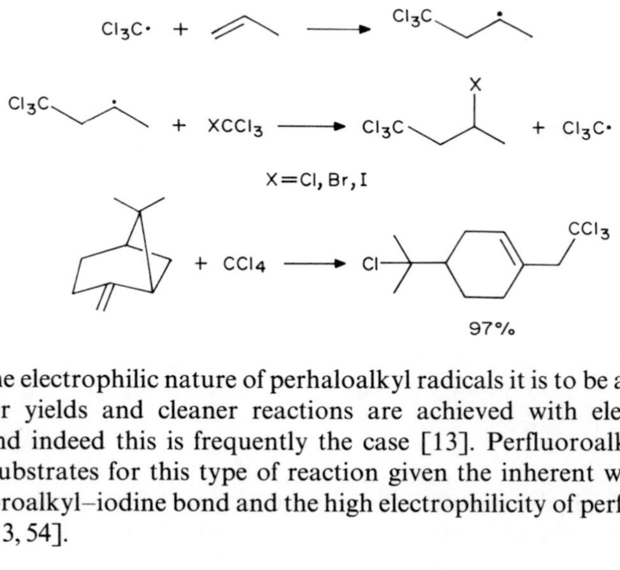

97%

Given the electrophilic nature of perhaloalkyl radicals it is to be anticipated that higher yields and cleaner reactions are achieved with electron rich alkenes, and indeed this is frequently the case [13]. Perfluoroalkyl iodides are ideal substrates for this type of reaction given the inherent weakness of the perfluoroalkyl–iodine bond and the high electrophilicity of perfluoroalkyl radicals [13, 54].

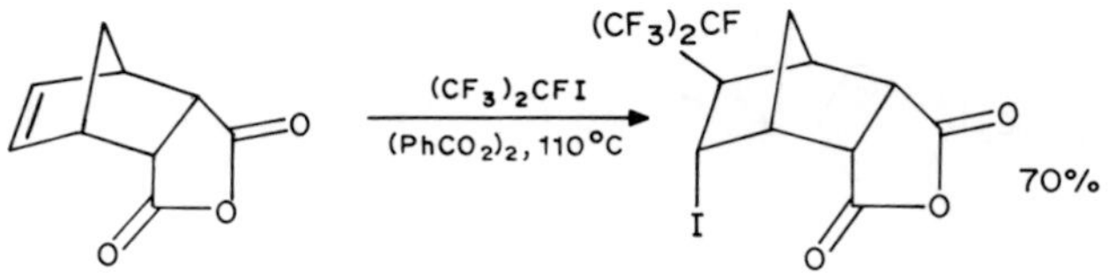

70%

The use of less reactive tetrachloromethane (stronger C—Cl bond) is advantageously catalysed by ferric chloride by an electron transfer chain process [55].

More recently, other electrophilic radicals derived from iodo- and bromo-esters and -nitriles have been successfully employed in this type of chemistry. Such reactions are reported to be highly efficient with bromomalononitrile [56].

$$BrCH(CN)_2 \; + \; \underset{}{\overset{}{\diagup\diagdown}} \xrightarrow{\; h\upsilon \;} (NC)_2CHC(Me)_2C(Me)_2Br$$

98%

Giese has applied this methodology successfully to bromomalonate esters. Extension to chloromalonates is, however, not possible as the strength of the carbon–chlorine bond effectively precludes chain transfer. In the presence of tri-n-butyltin hydride and electron rich alkenes, both bromo- and chloromalonates proceed via the normal three propagation step tin mediated addition reaction [57].

A synthetically useful procedure for the formation of butyrolactones by the radical addition of tri-n-butylstannyl α-iodoesters to electron rich alkenes followed by ring closure has been developed by Kraus and coworkers [58].

Formation of butyrolactones from tri-n-butylstannyl α-iodoesters and alkenes [59]

The appropriate ester (2 equiv.) and bis (tri-n-butyltin) oxide (1 equiv.) were heated to 130°C for 30 min. After cooling, the reaction was extracted with hot hexane. Removal of the hexane *in vacuo* provided the stannyl esters, which were used without purification.

The stannyl α-iodoester (1 equiv.), the alkene (3 equiv.) and AIBN (5 mol%) were heated to reflux in benzene (1 M solution) for 8 h. The reaction mixture was then partitioned between acetonitrile and hexane. Chromatography on silica gel of the acetonitrile phase provided the lactone.

Finally, a recent development of this halogen atom abstraction methodology involves the triethylborane mediated addition of alkyl iodides to enones. In this reaction sequence an ethyl radical abstracts iodine from the alkyl iodide giving an alkyl radical which undergoes addition to the enone. Chain transfer is closely related to that involved in the reaction of trialkylboranes with $\alpha\beta$-unsaturated carbonyl compounds and involves S_H2 at boron with liberation of a further ethyl radical. The borinyl enolates so obtained can be quenched by addition of methanol or reacted *in situ* with aldehydes in an aldol reaction [60].

6.5   ADDITION TO ALKENES WITH CONCOMITANT DISPLACEMENT OF A CHAIN CARRYING RADICAL

6.5.1   Distal addition–elimination procedures (S_H2' reactions)

The reaction of polyhalomethanes with allyltri-n-butyl- and allyltrimethyl-stannane via a two step chain sequence was first reported in 1973 [61] but has been little employed in organic synthesis.

A variation, introduced by Johnson [62], employs allylcobaloximes as the radical trap/source of propagating radical. With polyhalomethanes and haloesters these reactions proceed in high yield at room temperature with photochemical initiation [63].

71%

†[Co] = $C_5H_5N(dmgH)_2Co$; dmgH = dimethylglyoximato.

These reactions take place with allylic transposition and, importantly from a synthetic viewpoint, are relatively insensitive to substitution at the terminus of the allylcobaloxime [64].

Allenylcobaloximes have also been used in conjunction with polyhalo-methanes as sources of alkynes by a distal addition–elimination chain sequence [64].

It is only relatively recently, however, through the work of Keck, that this methodology has been extended to encompass nucleophilic alkyl radicals [65]. Alkyl halides and pseudohalides (phenylselenides, xanthates) are reacted with allyltri-n-butylstannane with AIBN initiation either in benzene or toluene at reflux or photochemically at room temperature. Many examples are now known of this extremely useful sequence which is compatible with the same spectrum of functionality and molecular complexity as the tin hydride methodology.

The reaction is readily extended to methallylstannanes [65].

72%

Substitution at the terminal position of the allylstannane, as in crotonyltri-n-butylstannane, is however not tolerated; hydrogen abstraction from an allylic position being an effective competing reaction [66].

Caution must also be observed with allylstannanes substituted α to tin as under the reaction conditions scrambling takes place [67].

*Reaction of allyltri-n-butylstannane with alkyl halides
(and pseudohalides)* [68]

Photochemical.   The alkyl bromide (1 mmol) and allyltri-n-butylstannane (2 mmol) were photolysed through Pyrex with a 100 W medium pressure mercury lamp in toluene (2 ml) at room temperature under an argon atmosphere until TLC control indicated that the reaction was complete (18–20 h). The solvent was removed *in vacuo* and the product isolated by chromatography on silica gel.

Thermal.   The alkyl bromide (10 mmol) and allyltri-n-butylstannane (20 ml) were heated to 80°C under argon in toluene (20 ml) with periodic addition of catalytic amounts of AIBN. After completion and removal of volatiles, the product was isolated by chromatography on silica gel.

The addition of nucleophilic radicals to the electron rich double bond of allyltri-n-butylstannane is a slow reaction leading to long reaction times requiring continual addition of initiator and is plagued by side reactions involving hydrogen atom abstraction from *inter alia* the allylic position of the reagent. Baldwin and coworkers have introduced a similar reagent in which the double bond is activated towards addition of nucleophilic radicals by a carboethoxy group [67].

70%

The reagent is conveniently prepared from 2-carboethoxyallyl bromide by displacement of bromide with sodium p-toluenesulphinate followed by reaction with tri-n-butyltin hydride [67, 69].

Related reagents have been developed by the Barton group for use in conjunction with the O-acyl thiohydroxamates [46, 70]. The leaving group in these reagents is a thiyl radical which propagates the chain by attack at the thiohydroxamate thiocarbonyl group in the usual manner.

Unlike the case of the allylstannane it was found that the presence of an activating group in the allyl thioether was essential for the obtention of good yields in this chemistry [70].

Carboethoxyallylation [71]

t-Butylmercaptan (0.4 ml) was added at room temperature to a vigorously stirred suspension of potassium carbonate (0.5 g) and ethyl 2,2′-dibromoiso-butyrate (1 g) [72] in absolute ethanol (5 ml). Stirring was maintained for 18 h, after which the reaction mixture was poured into water (100 ml) and extracted with ether (3 × 20 ml). Distillation of the crude product through a short Vigreux column gave ethyl 3-t-butyl-2-methylenepropionate (69%) as a colourless oil, b.p. 54°C/0.3 mmHg.

The appropriate acid chloride (1 mmol) in chlorobenzene (5 ml) was added over 5 min to a stirred suspension of N-hydroxypyridin-2-thione sodium salt (1.2 mmol) and the above reagent (2 mmol) in chlorobenzene (5 ml) at reflux under nitrogen. After completion (TLC + disappearance of yellow coloration) the reaction was cooled to room temperature, filtered on celite and the volatiles removed *in vacuo*. The product was isolated by chromatography on silica gel.

6.5.2 Proximal addition–elimination procedures

Carbon–carbon bond formation by a proximal addition–elimination procedure amounts to the *ipso* substitution of a radical leaving group by a carbon centred radical. A comprehensive survey of this area including a wide range of leaving groups and stereochemical aspects has been carried out by the Russell group [73].

In general, it is found that using E-alkenes leads largely to retention of configuration. With Z-alkenes low stereospecificities are observed. Stereospecificities are highest when the group Y is an efficient radical leaving group such that elimination occurs before free rotation takes place in the intermediate radical.

Baldwin has developed this chemistry into a synthetically useful method for the formation of carbon–carbon bonds using alkyl bromides or pseudo-halides as the source of the carbon radical [74].

74%

70%

6.6 ADDITION TO ALKYNES

Relatively little use has been made of alkynes in intermolecular radical carbon–carbon bond forming reactions. Nevertheless the same principles apply as for addition to alkenes; that is, nucleophilic radicals add preferentially to electron poor alkynes and electrophilic radicals add to electron rich alkynes.

The methodology available for radical addition to alkynes mirrors that in use for addition to alkenes and is not therefore discussed extensively here. From a stereochemical viewpoint, addition to singly and doubly substituted alkynes usually leads to mixtures of *cis*- and *trans*-alkenes (*syn* and *anti* addition, respectively). Giese and coworkers have conducted a study, using the mercury method of *syn* and *anti* addition to alkynes, and report that *syn* addition is the preferred mode except for the case of bulky incoming radicals such as ^{t}Bu [75].

Table 6.5 gives representative examples of intermolecular radical addition to alkynes by a variety of different methods.

Photochemically initiated addition of aldehydes and formamides to alkynes normally results in the formation of 2:1 adducts [79]

Finally, distal addition–elimination procedures involving alkynes substituted at the propargylic position and resulting in the formation of allenes have been devised [64, 80].

TABLE 6.5

Radical addition to alkynes

Radical source	Alkyne	Conditions	Product	Yield (%)	Ref.
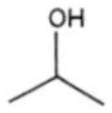 (isopropanol)	$\equiv$—CO_2H	$hv_3(Ph_2CO)$	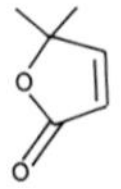 (furanone structure)	36	[76]
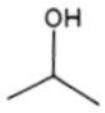 (isopropanol)	HO_2C—$\equiv$—CO_2H	hv	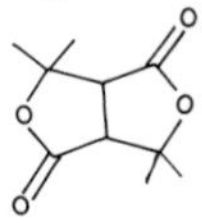 (bicyclic dione structure)	26	[76]
$^nC_6H_{13}HgX$	$\equiv CO_2Me$	$NaBH_4$	$^nC_6H_{13}CH$=$CHCO_2Me$ (c:t = 32:69)	10	[75]
$^cC_6H_{11}HgX$	$\equiv$—CO_2Me	$NaBH_4$	$^cC_6H_{11}CH$=$CHCO_2Me$ (c:t = 44:56)	35	[75]
$^tC_4H_9HgX$	$\equiv CO_2Me$	$NaBH_4$	$^tC_4H_9CH$=$CHCO_2Me$ (c:t = 72:28)	41	[75]
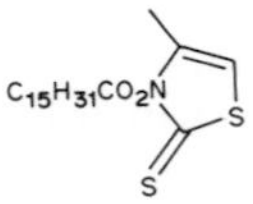 $C_{15}H_{31}CO_2N$ (thiazole thione)	$\equiv CO_2Me$	110°C	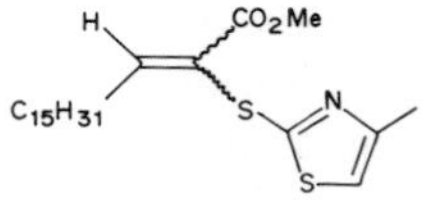 (vinyl sulfide product)	38	[45]
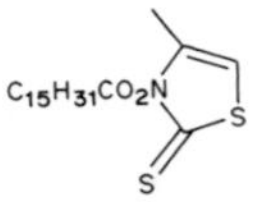 $C_{15}H_{31}CO_2N$ (thiazole thione)	MeO_2CC≡CCO_2Me	110°C	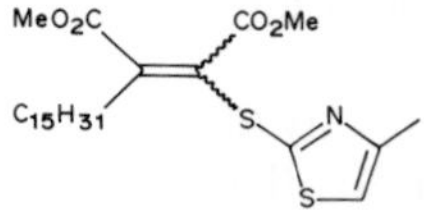 (vinyl sulfide product)	50	[45]
$CBrCl_3$	Ph—$\equiv$	hv	Cl_3CCH=$CBrPh$	32	[77]
CF_3I	HC≡CH	hv	CF_3CH=CHI	78	[78]

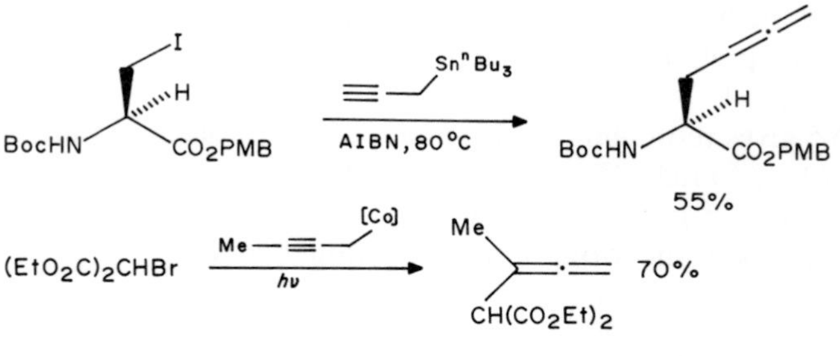

6.7   CARBON–CARBON BOND FORMATION BY HOMOLYTIC SUBSTITUTION AT AN sp³ CARBON

The formation of carbon–carbon bonds by the attack of a carbon centred radical at an sp³ carbon with concomitant displacement of a radical leaving group (S_H2 at carbon) is a rarely encountered process. One of the main reasons for which is the competing pathway of abstraction of the "leaving group" by the attacking radical.

The reaction of benzylcobaloximes with bromotrichloromethane leading to the formation of β,β,β-trichloroethylarenes via a radical chain mechanism is one example of such a process [62, 81].

$$PhCH_2-[Co] + BrCCl_3 \xrightarrow{h\nu} PhCH_2CCl_3 + Br-[Co]$$

More recently, [1.1.1]propellane has been shown to react with a variety of carbon centred radicals by a chain process involving attack of a carbon radical at a bridgehead position with cleavage of the central carbon–carbon bond [82]. Given the readily availability of [1.1.1]propellane, this method shows considerable promise for the preparation of bridgehead substituted bicyclo[1.1.1]pentanes.

REFERENCES

1. M. J. S. Dewar and S. Olivella, *J. Am. Chem. Soc.* **100**, 5290 (1978); K. N. Houk, M. N. Paddon-Row, D. C. Spellmeyer, N. R. Rondan and S. Nagase, *J. Org. Chem.* **51**, 2874 (1986).
2. For a detailed discussion of factors affecting radical addition to alkenes and alkynes see: B. Giese, *Angew. Chem. Int. Ed. Engl.* **22**, 753 (1983).
3. J. A. Baban and B. P. Roberts, *J. Chem. Soc. Perkin Trans. 2*, 161 (1981).
4. For a review on stereochemical aspects of radical reactions see: B. Giese, *Angew. Chem. Int. Ed. Engl.* **28**, 969 (1989).
5. N. A. Porter, D. M. Scott, B. Lacher, B. Giese, H. G. Zeitz and H. J. Lindner, *J. Am. Chem. Soc.* **111**, 8311 (1989); for a comparison with conformationally less rigid esters see: D. Crich and J. W. Davies, *Tetrahedron Lett.* **28**, 4205 (1987).
6. B. Giese, H. Harnisch and U. Lüning, *Chem. Ber.* **118**, 1345 (1985).
7. S. D. Burke, W. B. Fobare and D. M. Armistead, *J. Org. Chem.* **47**, 3348 (1982).
8. D. H. R. Barton, A. Gateau-Olesker, S. D. Gero, B. Lacher, C. Tachdjian and S. Z. Zard, *J. Chem. Soc. Chem. Commun.* 1790 (1987).

9. Y. Araki, T. Endo, M. Tanji, J. Nagasawa and Y. Ishido, *Tetrahedron Lett.* **29**, 351 (1988); D. H. R. Barton, S. D. Géro, B. Quicelet-Sire and M. Samadi, *J. Chem. Soc. Chem. Commun.* 1372 (1988).
10. J.-Y. Nedelec, D. Blanchet, D. Lefort and J. Guilhem, *J. Chem. Res.* (S) 315 (1987); for related examples with nucleophilic radicals see ref. [45].
11. B. Giese and J. Meixner, *Tetrahedron Lett.* 2783 (1977).
12. C. Walling, *Free Radicals in Solution.* Wiley, New York, 1957; E. S.Huyser, *Free Radical Chain Reactions.* Wiley Interscience, New York, 1970; G. Sosnovsky, *Free Radicals in Preparative Organic Chemistry.* MacMillan, London, 1964; C. J. M. Stirling, *Radicals in Organic Synthesis.* Oldbourne, London, 1965; D. Elad, in *Organic Photochemistry* (ed. O. L. Chapman), Vol. 2, p. 168. Dekker, New York, 1969.
13. C. Walling and E. S. Huyser, *Org. Reactions* **13**, 91 (1963).
14. Adapted from: B. Giese and J. Meister, *Chem. Ber.* **110**, 2588 (1977); G. Kretzschmar, Doctoral Thesis, Technischen Hochschule Darmstadt, Darmstadt, FRG, 1983.
15. *Houben-Weyl: Methoden der Organischen Chemie*, Vol.13/2b. Georg-Thiem Verlag, Stuttgart, 1974; R. C. Larock, *Organomercury Compounds in Organic Synthesis.* Springer-Verlag, Berlin, 1985.
16. D. Crich and J. W. Davies, *Tetrahedron* **45**, 5641 (1989).
17. B. Giese and K. Gröninger, *Tetrahedron Lett.* **25**, 2743 (1984).
18. Adapted from: B. Giese, K. Heuck, H. Lenhardt and Ü. Lüning, *Chem, Ber.* **117**, 2132 (1984).
19. B. Giese and K. Heuck, *Chem. Ber.* **114**, 1572 (1981).
20. B. Giese, T. Hasskerl and Ü. Lüning, *Chem. Ber.* **117**, 859 (1984).
21. A. P. Kozikowski and J. Scripko, *Tetrahedron Lett.* **24**, 2051 (1983).
22. R. Henning and H. Urbach, *Tetrahedron Lett.* **24**, 5343 (1983).
23. S. Danishefsky, E. Taniyama and R. R. Webb, *Tetrahedron Lett.* **24**, 11 (1983); S. Danishefsky and E. Taniyama, *Tetrahedron Lett.* **24**, 15 (1983).
24. B. Giese and W. Zwick, *Tetrahedron Lett.* **21**, 3569 (1980); B. Giese and W. Zwick, *Chem. Ber.* **112**, 3766 (1979).
25. B. Giese, H. Horler and W. Zwick, *Tetrahedron Lett.* **23**, 931 (1982); B. Giese and H. Horler, *Tetrahedron Lett.* **24**, 3221 (1983).
26. B. Giese and G. Kretzschmar, *Angew. Chem. Int. Ed. Engl.* **20**, 965 (1981).
27. B. Giese and J. A. González-Gómez, *Chem. Ber.* **119**, 1291 (1986).
28. D. Crich, J. W. Davies, G. Negrón and L. Quintero, *J. Chem. Res.* (S) 140 (1988).
29. Adapted from ref. [7].
30. R. M. Adlington, J. E. Baldwin, A. Basak and R. P. Kozyrod, *J. Chem. Soc. Chem. Commun.* 944 (1983).
31. B. Giese and J. Dupuis, *Angew. Chem. Int. Ed. Engl.* **22**, 622 (1983).
32. B. Giese and T. Witzel, *Angew. Chem. Int. Ed. Engl.* **25**, 450 (1986).
33. R. Blattner, R. J. Ferrier and R. Renner, *J. Chem. Soc. Chem. Commun.* 1007 (1987).
34. B.Giese, J. A. González-Gómez and T. Witzel, *Angew chem. Int. Ed. Engl.* **23**, 69 (1984).
35. N. Ono, H. Miyake, A. Kamimura, I. Hamamoto, R. Tamura and A. Kaji, *Tetrahedron* **41**, 4013 (1985).
36. G. Sacripante, C. Tan and G. Just, *Tetrahedron Lett.* **26**, 5643 (1985).
37. D. L. Boger and R. J. Mathvink, *J. Org. Chem.* **54**, 1777 (1989).
38. D. Crich and D. Batty, *Synthesis* 273 (1990).
39. P. Pike, S. Hershberger and J. Hershberger, *Tetrahedron Lett.* **26**, 6289 (1985); *Tetrahedron* **44**, 6295 (1988).
40. B. Giese, B. Kopping and C. Chatgilialoglu, *Tetrahedron Lett.* **30**, 681 (1989); M. Lesage, C. Chatgilialoglu and D. Griller, *Tetrahedron Lett.* **30**, 2733 (1989).
41. H. C. Brown and M. M. Midland, *Angew. Chem. Int. Ed. Engl.* **11**, 692 (1972).

REFERENCES 211

42. H. C. Brown and G. W. Kabalka, *J. Am. Chem. Soc.* **92**, 712, 714 (1970).
43. Adapted from ref. [42].
44. P. Jacob, *J. Organomet. Chem.* **156**, 101 (1978).
45. D. H. R. Barton, D. Crich and G. Kretzschmar, *J. Chem. Soc. Perkin Trans.* 1, 39 (1986).
46. D. H. R. Barton, H. Togo and S. Z. Zard, *Tetrahedron* **41**, 5507 (1985).
47. D. H. R. Barton, H. Togo and S. Z. Zard, *Tetrahedron Lett.* **26**, 6349 (1985).
48. D. H. R. Barton, Y. Hervé, P. Potier and J. Thierry, *Tetrahedron* **43**, 4297 (1987).
49. Adapted from ref. [45].
50. D. H. R. Barton, D. Bridon and S. Z. Zard, *Tetrahedron* **43**, 5307 (1987).
51. D. H. R. Barton, B. Garcia, H. Togo and S. Z. Zard, *Tetrahedron Lett.* **27**, 1327 (1986).
52. D. H. R. Barton, B. Lacher and S. Z. Zard, *Tetrahedron* **42**, 2325 (1986).
53. D. H.R. Barton and D. Crich. *J. Chem. Soc., Perkin Trans.* 1, 1603 (1986).
54. N. O. Brace, *J. Org. Chem.* **44**, 1964 (1979).
55. F. Minisci, *Acc. Chem. Res.* **8**, 165 (1975).
56. P. Boldt, L. Schultz and J. Etzemüller, *Chem. Ber.* **100**, 1281 (1967); M. H. Tredar, H. Kratzin, H. Lübbecke, C. Y. Yang and P. Boldt, *J. Chem. Res.* (S) 165 (1977).
57. B. Giese, H. Horler and M. Leising, *Chem. Ber.* **119**, 444 (1986).
58. G. A. Kraus and K. Landgrebe, *Tetrahedron* **41**, 4039 (1985); M. Degeuil-Castaing, B. de Jeso, G. A. Kraus, K. Landgrebe and B. Maillard, *Tetrahedron Lett.* **27**, 5927 (1986).
59. Adapted from ref. [58].
60. K. Nozaki, K. Oshima and K. Utimoto, *Tetrahedron Lett.* **29**, 1041 (1988).
61. M. Kosugi, K. Kurino, T. Takayama and T. Migata, *J. Organomet. Chem.* **56**, C11 (1973); J. Grignon, C. Servens and M. Pereyre, *J. Organomet. Chem.* **96**, 225 (1975).
62. M. D. Johnson. *Acc. Chem. Res.* **16**, 343 (1983).
63. M. Veber, K. N. V. Duong, F. Gaudemer and A. Gaudemer, *J. Organomet. Chem.* **177**, 231 (1979).
64. A. Bury, C. J. Cooksey, T. Funabiki, B. D. Gupta and M. D. Johnson, *J. Chem. Soc. Perkin Trans.* 2, 1050 (1979); A. Bury, S. T. Corker and M D. Johnson, *J. Chem. Soc. Perkin Trans.* 1, 645 (1982); M. Veber, K. N. V. Duong, A. Gaudemer and M. D. Johnson, *J. Organomet. Chem.* **209**, 393 (1981).
65. G. E. Keck, E. J. Enholm, J. B. Yates and M. R. Wiley, *Tetrahedron* **41**, 4079 (1985).
66. G. E. Keck and J. B. Yates, *J. Organomet. Chem.* **248**, C21 (1983); but see: G. E. Keck and J. H. Byers, *J. Org. Chem.* **50**, 5442 (1985) for an alternative procedure.
67. J. E. Baldwin, R. M. Adlington, D. J. Birch, J. A. Crawford and J. B. Sweeney, *J. Chem. Soc. Chem. Commun.* 1339 (1986).
68. Adapted from ref. [65].
69. J. Villieras and M. Rambaud, *Synthesis* 924 (1982).
70. D. H. R. Barton and D. Crich, *J. Chem. Soc. Perkin Trans.* 1, 1613 (1986).
71. Adapted from ref. [70].
72. A. F. Ferris, *J. Org. Chem.* **20**, 780 (1955).
73. G. A. Russell, H. Tashtoush and P. Ngoviwatchai, *J. Am. Chem. Soc.* **106**, 4622 (1984); G. A. Russell and P. Ngoviwatchai, *Tetrahedron Lett.* **26**, 4975 (1985); G. A. Russell, *Acc. Chem. Res.* **22**, 1 (1989).
74. J. E. Baldwin, D. R. Kelly and C. B. Ziegler, *J. Chem. Soc. Chem. Commun.* 133 (1984); J. E. Baldwin and D. R. Kelly, *J. Chem. Soc. Chem. Commun.* 682 (1985).
75. B. Giese and S. Lachhein, *Angew. Chem. Int. Ed. Engl.* **21**, 768 (1982); B. Giese, J. A. González-Gómez, S. Lachhein and J. O. Metzger, *Angew. Chem. Int. Ed. Engl.* **26**, 479 (1987).
76. M. Pfau, R. Dulou and M. Vilkas, *Compt. Rendu* **254**, 1817 (1962).
77. M. S. Kharasch and M. Sage, *J. Org. Chem.* **14**, 537 (1949).
78. R. N. Haszeldine, *J. Chem. Soc.* 3037 (1950); R. N. Haszeldine, *J. Chem. Soc.* 588 (1951).

79. G. Friedman and A. Komem, *Tetrahedron Lett.* 3357 (1968); R. H. Wiley and J. R. Harrell, *J. Org. Chem.* **25**, 903 (1960).
80. J. E. Baldwin, R. M. Adlington and A. Basak, *J. Chem. Soc., Chem. Commun.* 1284 (1984); G. A. Russell and L. L. Herold, *J. Org. Chem.* **50**, 1037 (1985).
81. However for an alternative mechanistic interpretation see: R. C. McHatton, J. H. Espenson and A. Bakac, *J. Am. Chem. Soc.* **108**, 5885 (1986).
82. K. B. Wiberg, S. T. Waddell and K. Laidig, *Tetrahedron Lett.* **27**, 1553 (1986); K. B. Wiberg and S. T. Waddell, *Tetrahedron Lett.* **28**, 151 (1987); J. Belzner and G. Szeimies, *Tetrahedron Lett.* **28**, 3099 (1987); P. Kaszynski and J. Michl, *J. Am. Chem. Soc.* **110**, 5225 (1988); K. B. Wiberg, *Chem. Rev.* **89**, 975 (1989).

−7−

Intramolecular Carbon–Carbon Bond Forming Free Radical Chain Reactions

7.1 PREPARATION OF FIVE MEMBERED RINGS BY 5-*EXO* CYCLIZATIONS

The kinetically controlled rearrangement of variously substituted 5-hexenyl radicals to substituted cyclopentylmethyl radicals constitutes an excellent method for the preparation of five membered rings. The high preference ($\simeq 98{:}2$) for cyclization in the *exo* mode leading to five membered rings over that for *endo* ring closure (six membered ring formation) is best understood in terms of a relatively strain free chairlike transition state accommodating the stereoelectronic requirements of radical additional to double bonds [1].

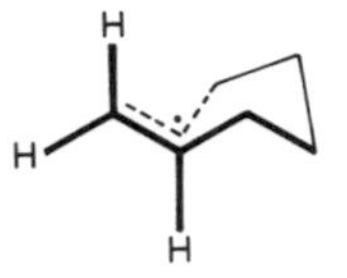

This model also provides a satisfactory rationale for the observed stereoselectivities in the cyclization of 2,3- and 4-substituted hexenyl radicals. Thus 2- and 4-methyl-5-hexenyl cyclizations occur with preferential formation of the *trans* rather than *cis* disubstituted cyclopentanes whereas the 3-methyl-5-hexenyl radical gives a 3:1 ratio of *cis/trans*-dimethylcyclopentane under similar conditions [2]. These observations point to the preferred adoption of pseudoequatorial positions by the substituents in the respective transition states.

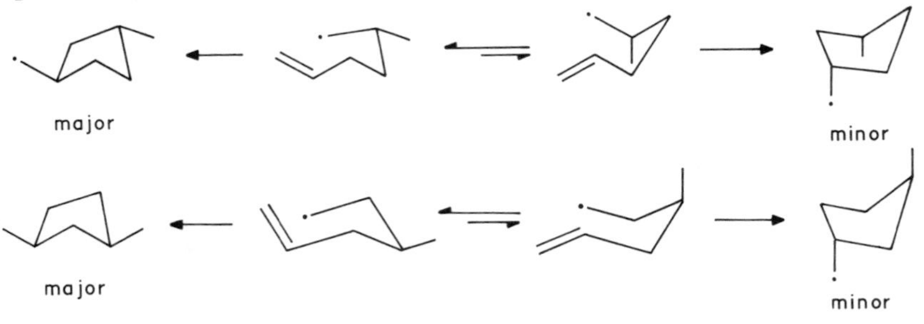

major   minor

major   minor

major   minor

In the case of the 1-methyl-5-hexenyl radical, the major product is
cis-1,2-dimethylcyclopentane [3]. The reasons for this preference may be
steric or stereoelectronic [1,3].

major   minor

This preference for the formation of cis-1,2-disubstituted systems from
1-substituted 5-hexenyl radicals extends to the formation of bicyclic species
from 2-(3-butenyl)cycloalkyl radicals. Thus, the 2-(3-butenyl)cyclopentyl
radical cyclizes to give an 8:1 ratio of the endo- and exo-[3.3.0]bicyclo-
octylmethyl radicals [4].

exo-   endo

1   8

Calculations [1] and experiment [5], however, suggest that increasing the
bulk of the substituent at the 1-position leads to the preferential formation
of the trans product.

Indeed, RajanBabu has presented more complex examples in which very
high trans selectivity about the newly formed bond was found [6].

50%

7.1.1 The effect of substituents on the rate and regioselectivity of 5-hexenyl cyclizations

Table 7.1 illustrates the effect of methyl substituents on the rate of *exo* and *endo* ring closure of 5-hexenyl radicals. With the exception of the 5-methyl group, substituents generally serve to accelerate the ring closure reaction. A substituent in the 5-position sufficiently retards attack at that position to enable the *endo* mode to become an effective competing reaction. In practice, however, the more problematic competing reaction is quenching of the open chain radical.

TABLE 7.1

The effect of substituents on 5-hexenyl cyclizations

Radical	k_{25} (exo)	k_{25} (endo)
	2.3×10^5	4.1×10^3
	5.3×10^3	9.0×10^3
	3.5×10^5	6.0×10^3
	3.6×10^6	$<1 \times 10^5$
	5.1×10^6	$<1 \times 10^5$
	3.2×10^6	$<1 \times 10^5$

Adapted from A. L. J. Beckwith and C. H. Schiesser, *Tetrahedron* **41**, 3925 (1985).

7.1.2 Formation of carbocycles by 5-hexenyl cyclizations

The literature is replete with examples of 5-hexenyl type cyclizations in organic synthesis. The following examples are intended to illustrate the range of functional groups and methods of radical generation compatible with such cyclization reactions.

Some of the earliest and simplest examples were carried out by Brace and make use of the addition of electrophilic radicals to 1,6-heptadienes with chain transfer by halogen abstraction [7].

TABLE 7.2

Tin hydride mediated 5-hexenyl cyclizations

Substrate	Conditions	Product	Yield (%)	Ref.
	^{n}Bu$_3$SnH, 80°C		80	[8]
	^{n}Bu$_3$SnH, 80°C		71, 14	[9]
	^{n}Bu$_3$SnH		98	[10]
	Ph$_3$SnH, 80°C		58	[11]
	^{n}Bu$_3$SnH, 60°C		81	[12]
	^{n}Bu$_3$SnH, 80°C		62	[13]
	^{n}Bu$_3$SnH, 110°C		75	[14]

TABLE 7.2 *Continued*

Substrate	Conditions	Product	Yield (%)	Ref.
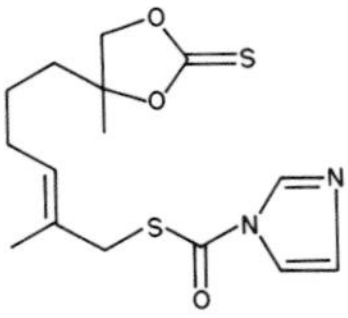	^{n}Bu$_3$SnH, 110°C	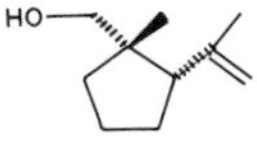	98	[15]
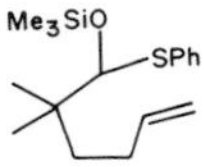	^{n}Bu$_3$SnH	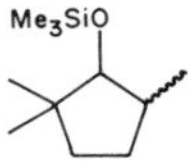 *trans:cis* ≃ 2:1	75	[16]
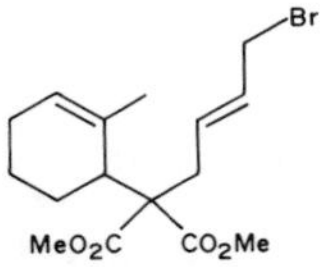	^{n}Bu$_3$SnH, 80°C	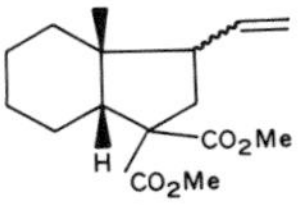 $\beta{:}\alpha = 3{:}1$	92	[17]

The tin hydride method of radical generation has been used extensively (Table 7.2).

The mercury and thiohydroxamate methods have also been employed, though much less extensively.

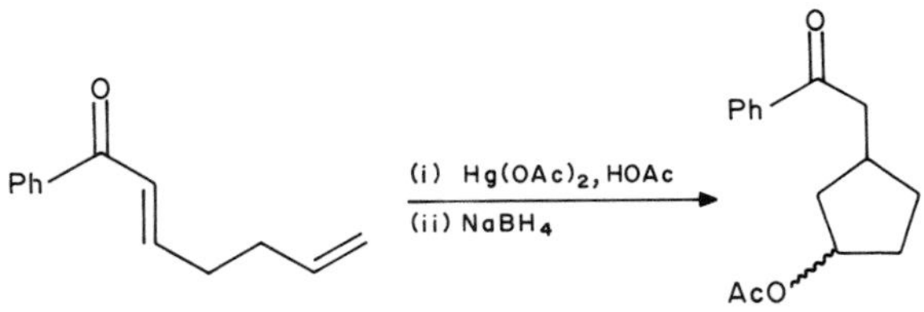

[18]

70%

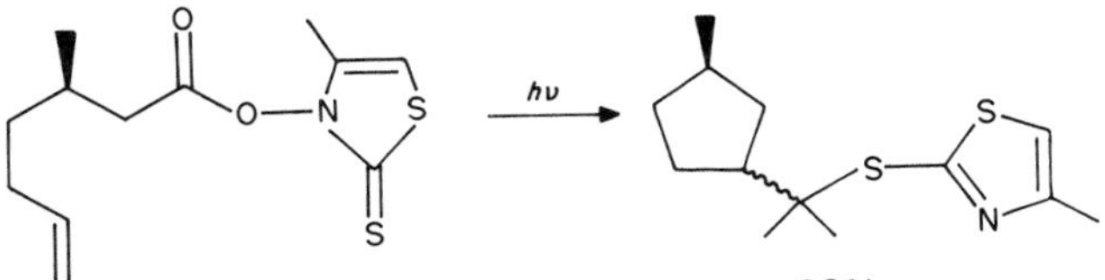

[19]

82%

The iodine atom transfer method has also been employed in cyclizations [20, 21].

7.1.3 Heterocycle synthesis by 5-hexenyl type cyclizations

The replacement of a methylene group of the 5-hexenyl chain with an oxygen atom or an amino group permits the ready synthesis of saturated five-membered heterocycles by radical cyclizations (Table 7.3).

The inclusion of ester and amide linkages into the "5-hexenyl" chain can be problematic. Thus, cyclizations of $\alpha\beta$-unsaturated esters and amides with the radical precursor in the O- or N-chain proceed well, even to the extent

TABLE 7.3

Heterocycles by hetero-5-hexenyl cyclizations

Substrate	Conditions	Product	Yield (%)	Ref.
	$CF_3CF_2CF_2I$, 60°C		80	[22]
	NaI, $^nBu_3SnH$, DME		70	[23]
	$^nBu_3SnH$, 80°C		87	[24]

TABLE 7.3 *Continued*

Substrate	Conditions	Product	Yield (%)	Ref.
	^{n}Bu$_3$SnH, 80°C	7 : 3	85	[25]
	hv		68	[26]
	^{n}Bu$_3$SnH, 80°C		67	[27]
	^{n}Bu$_3$SnH, 80°C		79	[28]
	^{n}Bu$_3$SnH, 110°C		65	[29]
	^{n}Bu$_3$SnH		88	[30]
	^{n}Bu$_3$SnH, 70°C		79 > 95 *trans*	[31]
	CCl$_4$, (PhCO$_2$)$_2$	*exo:endo* = 5:2	63	[32]

of overcoming the unfavourable polarization of the double bond in some cases.

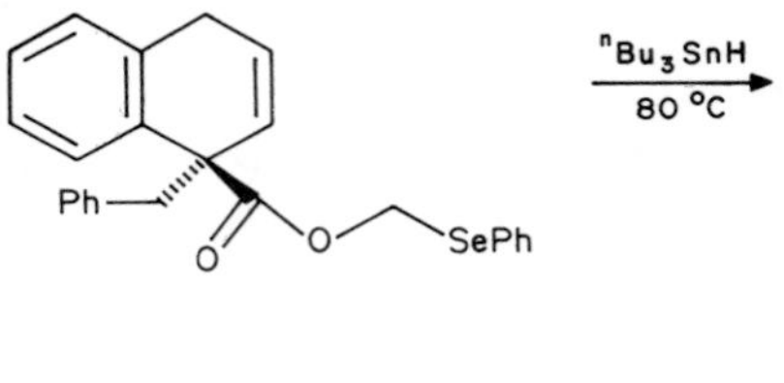

[33]

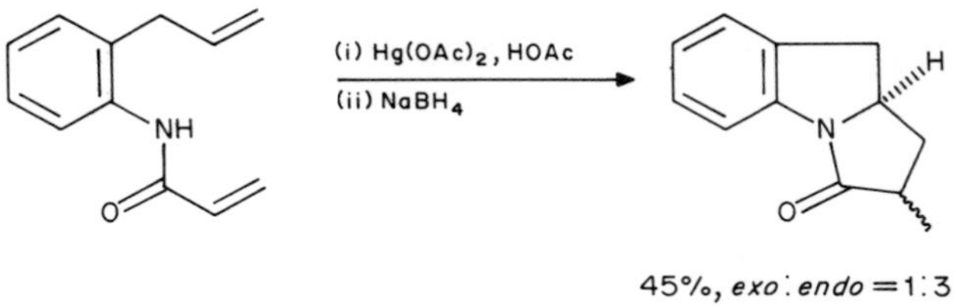

65%

[34]

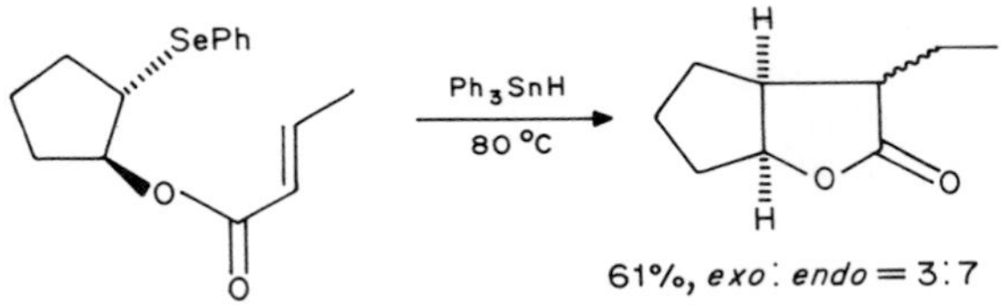

45%, exo : endo = 1 : 3

[35]

61%, exo : endo = 3 : 7

Whilst cyclizations of allyl esters and allyl amides of α-bromoacids proceed in low yield if at all owing to the *trans* configuration adopted by esters and amides.

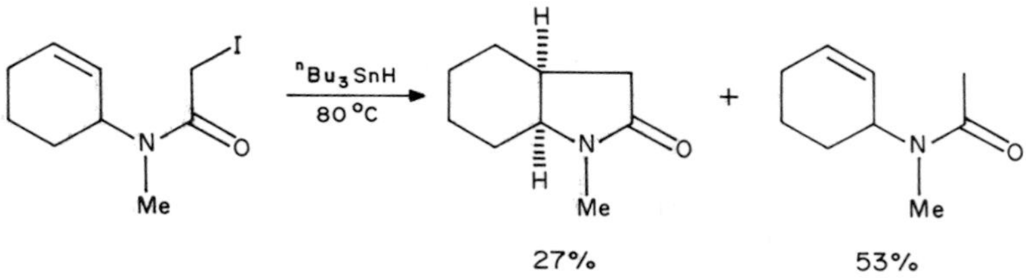

[36]

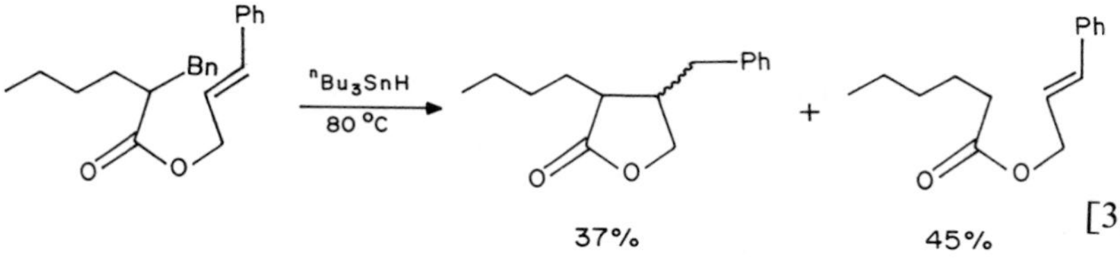

27%   53%

[37]

37%   45%

Several procedures have been developed to overcome this problem, the earliest and most widely used being the Stork bromoacetal method in which the latent ester function is revealed, after cyclization, by Jones oxidation [38].

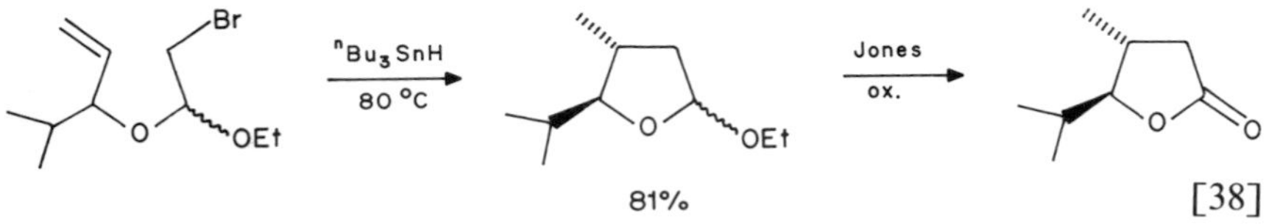

81%   [38]

[39]

95%

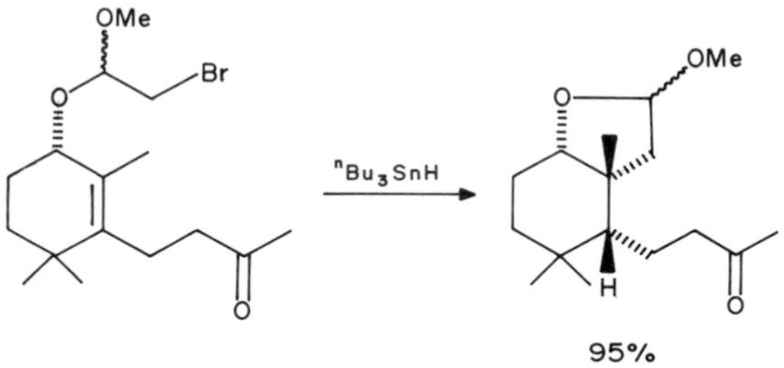

Alternative work up procedures enable the overall introduction of a methyl group [40].

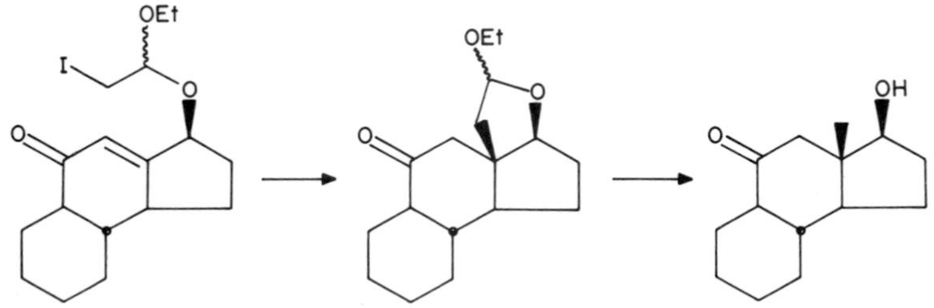

Stork [41] has also described a solution to the amide problem, in which the amide is temporarily converted into a mixed trifluoroacetimide or toluenesulphonamide.

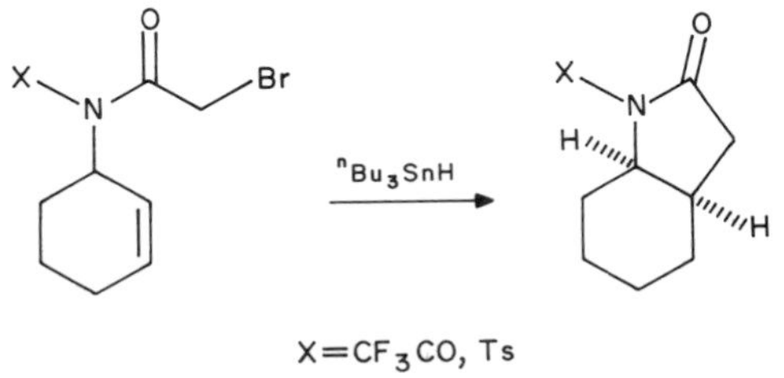

X=CF₃CO, Ts

An equally elegant procedure was devised by Livinghouse in which ethyl iodide is used as an addition source of iodide in an atom transfer cyclization. Radicals which adopt the wrong conformation for cyclization are quenched by iodine transfer and are subsequently available for recycling [36].

The attempted fusion of five-membered rings onto β- and γ-lactams by 5-hexenyl type cyclizations of aminomethyl radicals results in competing, and in the case of β-lactams exclusive, *endo* mode cyclization, due almost certainly to the reversibility of the ring closure and the greater strain in the *exo* mode products.

The extent of *endo* cyclization can be reduced, at least for γ-lactams [43], by substitution at the terminal position of the alkene. Alternatively, cyclization may be directed to the *exo* mode by the inclusion of an electron withdrawing group at the terminus of the alkene [43, 44] or by trapping of the kinetic *exo* mode radical by a β-elimination reaction [45].

Relatively little synthetic work has been carried out with other heteroatom substituted 5-hexenyl-like systems. Japanese workers [46] and Stork [47] have demonstrated the utility of bromomethyldimethylsilyl ethers of allylic alcohols in 5-hexenyl type cyclizations. After cyclization the system can be manipulated to give the product of effective reductive alkylation. When a fused bicyclic system is generated in the cyclization, high stereoselectivity is obtained.

The use of allylic alcohols substituted at the internal position with such a system leads, in good yield, to six-membered rings by *endo* cyclization [48].

It should also be noted that simple 2- and 3-(dimethylsila)-5-hexenyl radicals undergo preferential *endo* mode cyclization but that the 4-(dimethylsila)-5-hexenyl radical prefers the *exo* mode [49].

7.1.4 5-Hexynyl type cyclizations

Although somewhat slower $(k_{exo}(25°C) = 2.8 \times 10^4 \, s^{-1}; \; k_{endo}(25°C) < 6 \times 10^2 \, s^{-1})$ than the 5-hexenyl cyclization, the 5-hexynyl cyclization has been extensively used in organic synthesis. This type of cyclization possesses two advantages over the hexenyl type cyclization in so far as no new chiral centre is generated in the reaction and the newly formed *exo* cyclic double bond may be further manipulated. Representative examples are given in Table 7.4.

TABLE 7.4

5-Hexynyl cyclizations

Substrate	Conditions	Product	Yield (%)	Ref.
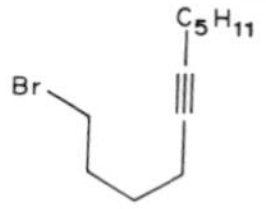	$^nBu_3SnH$, 80°C	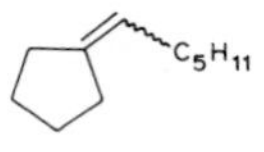	99	[50]
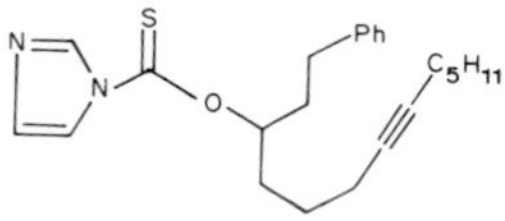	Ph_3SnH, 80°C	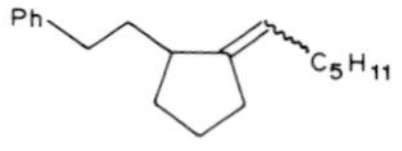	81	[51]
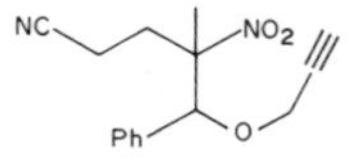	$^nBu_3SnH$, 80°C	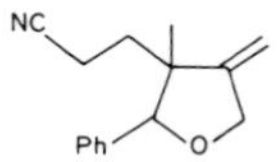	79	[25]
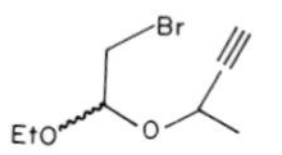	$^nBu_3SnH$, 80°C	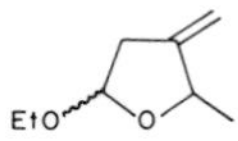	60	[52]
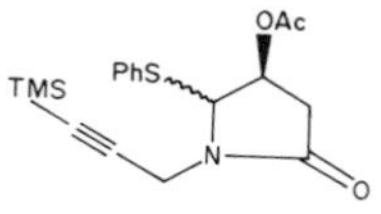	$^nBu_3SnH$, 80°C	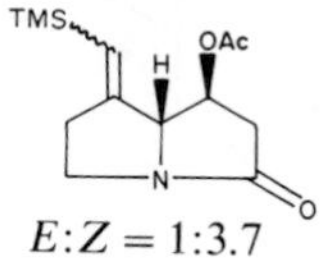 $E:Z = 1:3.7$	71	[53]

TABLE 7.4 *Continued*

Substrate	Conditions	Product	Yield (%)	Ref.
	Ph_3SnH, 80°C		86	[54]
	$^nBu_3SnH$, 80°C		85	[55]
	$^nBu_3SnH$, 80°C		85	[44]
	$^nBu_3SnH$, 80°C		88	[56]
	$(C_6H_{11})_3SnH$, 80°C		40	[57]
			8	
	$(PhCO_2)_2$, 80°C		47	[58]
	$^nBu_3SnH$, 110°C		95	[59]

7.1.5 Formation of five membered rings by cyclization onto allenes

Allenes may be used in 5-hexenyl type cyclizations, although the number of examples is relatively few. Attack takes place on the sp centre to generate, after ring closure, an allyl radical leading after chain transfer to mixtures of *endo* and *exo* cyclic regioisomers.

[60]

45%   51%

[43]

52%   14%

7.1.6 Experimental procedures for 5-hexenyl, 5-hexynyl and 4,5-hexadienyl cyclizations

The tin hydride chain is by far the most commonly used method for 5-hexenyl and related cyclizations. A wide variety of experimental methods are reported in the literature. A good reliable method is the dropwise addition of the tin hydride and a catalytic amount of AIBN ($\simeq 5 \text{ mol} \%$) in benzene to a solution of the radical precursor at reflux in benzene under nitrogen, much as described in Chapter 3 for the Barton–McCombie reaction. In those cases where cyclization is slow it is essential to maintain the tin hydride concentration in the reaction vessel as low as is compatible with chain propagation, and this is best achieved by motor driven syringe pump addition.

For the mercury and thiohydroxamate methods a good guideline is to use the same procedure as for the simple functional group transformations described in Chapter 3.

Preparation of diethyl 3-chloromethyl-4-(2,2,2-trichloroethyl)cyclopentyl)-1,1-dicarboxylate [7]

Tetrachloromethane (77 g, 0.50 mol), diethyl 2,2-diallylmalonate (24.0 g, 0.10 mol) and a solution of acetonitrile (8.2 g, 0.20 mol), ferric chloride hexahydrate (0.27 g, 1 mmol), benzoin (0.21 g, 2 mmol) and diethylamine hydrochloride (0.16 g, 1.5 mmol) were heated to 77°C under an atmosphere of carbon dioxide (nitrogen will suffice) for 22 h. The reaction mixture was washed with hydrochloric acid (5%, 25 ml) and dried (MgSO$_4$). Short path distillation (bath temperature, 192°C) gave the title product b.p. 149–155°C/0.15–0.25 mmHg.

Preparation of 1-bromo-2-methylbicyclo[3.1.0]hexane [9]

Tri-n-butyltin hydride (0.4 mol) and AIBN (1.7 mol%) in benzene (1000 ml) were added dropwise over 6 h to a solution of 1,1-dibromo-2-(4-butenyl) cyclopropane (0.35 mol) at reflux in benzene (1000 ml). After a further 3 h at reflux the benzene was carefully distilled off and the residue fractionated to separate the products from the tri-n-butyltin bromide.

Preparation of 9-methyl-3-oxo-2-azatricyclo[6.2.1.0$^{2.6}$]undecane [14]

To a solution of tri-n-butyltin hydride (304 mg, 1.04 mmol) in toluene (6 ml) at reflux under argon was added *O*-(9-methyl-3-oxo-2-azatricyclo[6.2.1.0$^{2.6}$] undecyl)-*S*-methyl dithiocarbonate (192 mg, 0.67 mmol) in toluene (6 ml) over 50 min. The solution was then heated to reflux for a further 17 h then concentrated *in vacuo* and purified by chromatography on silica gel (eluant: hexane/ethyl acetate 3:7) to give the title product as an oil (89 mg, 74%) with b.p. 60–70°C/0.14 mmHg (bulb to bulb) as a 2:1 mixture of isomers.

Preparation of 1,1-dimethyl(3-methylcyclopentyl)methyl
2-(4-methyl)thiazolylsulphide

(*R*)-Citronellic acid (1.07 g, 10.0 mmol) was dissolved in benzene (20 ml) and treated with oxalyl chloride (4.0 g, 30 mmol) and dimethyl formamide (1 drop). The mixture was stirred for 40 min at room temperature and then evaporated to dryness and the residue taken up in ether (10 ml); the latter was then added to a solution of 3-hydroxy-4-methylthiazole-2-thione (1.50 g, 10.2 mmol) in ether (40 ml) and pyridine. The mixture was stirred for 10 min at room temperature after which it was filtered, concentrated and then filtered through silica gel using ether–pentane as eluant to yield the *O*-acyl thiohydroxamate as a yellow oil (2.94, 98%). A solution of this ester (1.55 g, 5.18 mmol) in ether (20 mmol) was irradiated under nitrogen at room temperature with a 100 W medium pressure mercury lamp for 45 min. The solvent was then evaporated and the residue chromatographed on silica gel with pentane–ether (95:5) as the eluant to give the title product as a mixture of diastereoisomers (1.08 g, 82%) in the form of a colourless oil.

From ref. [19] with permission.

General procedure for the formation of γ-lactones from
stannyl α-iodocarboxylates

The requisite acid (2 equiv.) and bis(tri-n-butyltin) oxide (1 equiv.) were heated to 130°C for 30 min. The mixture was allowed to cool and then extracted with hot hexane. The hexane was removed *in vacuo* to afford the stannyl ester which was used as such. The stannyl iodocarboxylate in benzene (1 M solution) with 5 mol% of AIBN was then heated to reflux for 8 h. After removal of the benzene the reaction mixture was partitioned between hexane and acetonitrile and the product isolated from the acetonitrile phase and purified by chromatography on silica.

From ref. [20] with permission.

General photolytic procedure for iodine atom transfer cyclization [21]

The iodide was dissolved in benzene (0.3 M) and hexa-n-butylditin (0.07–0.1 equiv.) was added. The solution, under an inert atmosphere, was placed 6–10 cm from a 275 W sunlamp and irradiated for 5–30 min after which the reaction temperature was approximately 60–80°C. For less reactive substrates an additional portion of ditin was then added and irradiation continued. Appearance of the characteristic iodine colour indicated that the ditin had been consumed. The solvent was then evaporated and the crude reaction mixture treated with DBU (see Section 1.7.4) before purification by chromatography.

Standard method for the preparation and cyclization of bromoacetals [38]

Bromoacetals were prepared by the reaction of the appropriate allylic alcohol with a vinyl ether and N-bromosuccinimide, using the vinyl ether as the solvent at − 20 to − 50°C for 2 h. Tri-n-butyltin hydride (1 equiv.) in benzene (0.25 M) was then added dropwise to a solution of the bromoacetal and 1 mol % AIBN in benzene at 50°C. After stirring for a further 2 h the products were isolated by chromatography.

Preparation of 1,3-diols by cyclization of allyl
bromomethyl(dimethyl)silyl ethers

Silyl ethers were prepared by treatment of allylic alcohols with bromo-methyl(dimethyl)silyl chloride and triethylamine in dichloromethane at room temperature for several hours. To a solution of the silyl ether (2 mmol) in benzene (36 ml) at reflux under nitrogen was added dropwise over 2 h a solution of tri-n-butyltin hydride (2.2–2.5 mmol) and AIBN (0.03 mmol) in benzene (4 ml). Heating was continued for a further 1–2 h. The benzene was then evaporated under reduced pressure (unless the product was volatile) and the product treated with 30% hydrogen peroxide (1.2 ml) and potassium fluoride (10 mmol) in dimethyl formamide at 60°C for 7–8 h or in a mixture of methanol (3 ml) and tetrahydrofuran (3 ml) containing sodium carbonate (2 mmol) for 5 h at reflux. After standard work up, purification was by chromatography on silica.

From ref. [46] with permission.

7.1.7 Vinyl radical cyclizations

Extensive use has been made in organic synthesis of the cyclization of vinyl radicals onto alkenes in the 5-*exo*-trig mode. The methylenecyclopentanes generated in this manner are often accompanied by methylenecyclohexanes resulting from *overall endo* mode cyclization. Under appropriate conditions the latter product may dominate.

The formation of the *endo* mode products is rationalized by kinetic cyclization in the *exo* mode followed by a rapid rearrangement [62, 63, 71].

Vinyl radical cyclizations of this type are fast, and it has been demonstrated that the ring expansion (cyclohexane formation) can be virtually eliminated by operating with a high concentration of stannane or even in the limit in neat stannane without significant formation of the reduction (ring open) product [62]. Representative examples of this class of cyclization, whose main advantage is the retention of functionality in the form of an *exo* cyclic double bond are given in Table 7.5. It should be noted that the stereochemistry (if any) of the initial vinyl halide is of no consequence owing to the facile inversion of vinyl radicals.

It has been recently demonstrated how vinyl radical cyclization may be directed to give a high preference for *endo* cyclization by substituting the internal position of the alkene with radical stabilizing groups [65, 70].

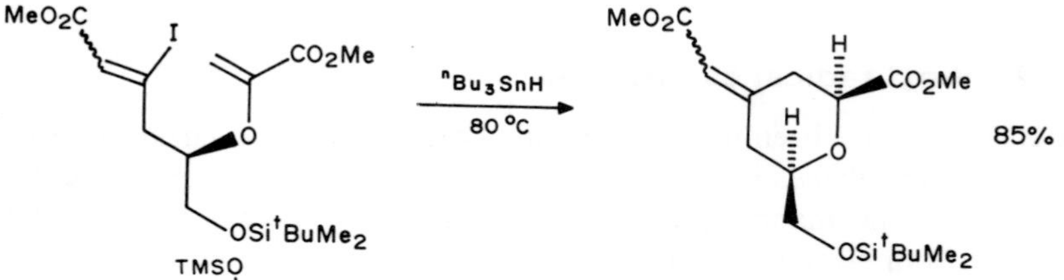

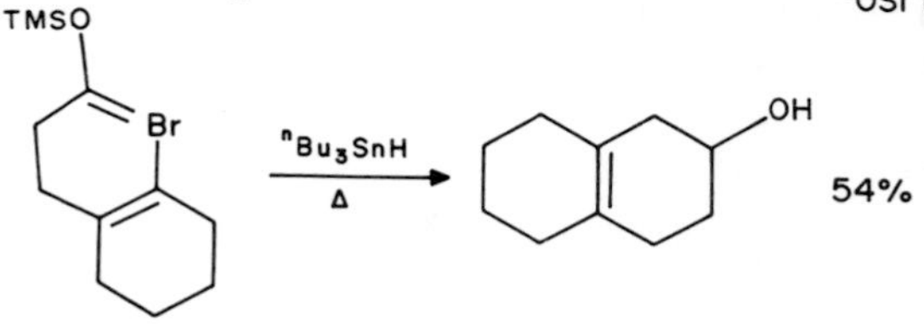

TABLE 7.5

Vinyl radical cyclizations

Substrate	Conditions	Product (Yield %)	Ref.
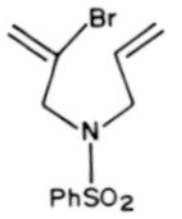	$^nBu_3SnH$, 80°C	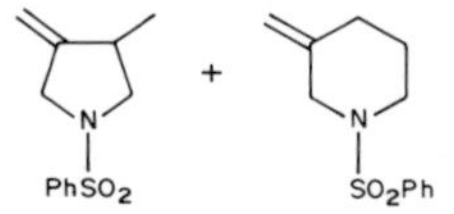 (24) + (49)	[24]
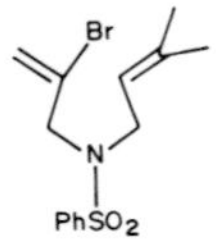	$^nBu_3SnH$, 80°C	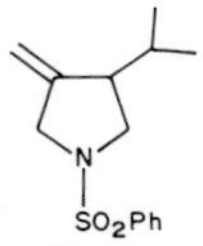 (60)	[24]
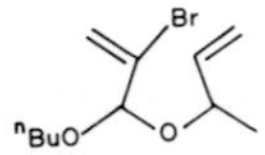	$^nBu_3SnH$, 80°C	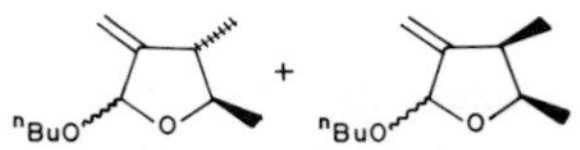 (63) + (2)	[64]

TABLE 7.5 *Continued*

Substrate	Conditions	Product (Yield %)	Ref.
	$^nBu_3SnH$, 80°C	(85)	[65]
	$^nBu_3SnH$, 80°C + hv	(70)	[61]
	$^nBu_3SnH$, 80°C	(86)	[66]
	$^nBu_3SnH$, 80°C	(47) + (24)	[61]
	$^nBu_3SnH$, 80°C	(12) + (61)	[67]
	$^nBu_3SnH$, hv	(68) + (10) $\beta:\alpha = 6:1$	[68]
	$^nBu_3SnH$, 80°C	(80)	[69]

General procedure for vinyl radical cyclizations

The vinyl iodide or bromide (200 mg) in benzene (34 ml) containing AIBN (2 mg) and tri-n-butyltin hydride (1.1 equiv.) was irradiated with a 275 W sunlamp. Heat from the sunlamp was used to keep the solution at reflux (0.5–1 h for iodides and 3–4 h for bromides). After removal of benzene the residue was stirred rapidly for 1 h with ether (5 ml) and saturated aqueous potassium fluoride (5 ml). Standard work up and chromatography on silica provided the cyclization products.

From ref. [61] with permission.

Stork has developed a method for vinyl radical cyclization, relying on the *reversible* addition of stannyl radicals to alkenes and alkynes [71], in which

TABLE 7.6

β-Stannylvinyl radical cyclizations

Substrate	Conditions	Product	Yield (%)	Ref.
	^{n}Bu$_3$SnH, 80°C		85	[71]
	^{n}Bu$_3$SnH, 80°C		70	[71]
	Ph$_3$SnH, Et$_3$B cat.		84	[73]
	^{n}Bu$_3$SnH, 110°C		50	[74]
	Ph$_3$SnH, 80°C		40	[75]

a vinyl radical is prepared by the addition of a stannyl radical to an appropriately placed triple bond (Table 7.6). The stannyl group is conveniently removed after the cyclization by protodesilylation with silica gel in dichloromethane or by transmetallation with butyl-lithium and subsequent quenching. An experimental procedure for this type of reaction is available in *Organic Synthesis* [72].

An interesting variant involving sulphonyl radical addition to a triple bond followed by cyclization of the so formed vinyl radical and eventual β-elimination of a second chain carrying sulphonyl radical has been described [32].

7.1.8 Aryl radical cyclizations

The very high rates of cyclization of 2-(3-butenyl)phenyl and related radicals has made aryl radical cyclizations popular methods for the preparation of benzo fused five membered rings.

		$k_{25}(s^{-1})$
$X = CH_2$,	$Y = H$	3.1×10^8
$X = O$,	$Y = H$	5.3×10^9
$X = O$,	$Y = Me$	1.7×10^9

Representative examples of this class of cyclization are given in Table 7.7.
Aryl radical cyclizations may be directed to the *endo* mode by the inclusion of an appropriate radical stabilizing group at the internal alkene position[65].

TABLE 7.7

Aryl radical cyclizations

Substrate	Conditions	Product	Yield (%)	Ref.
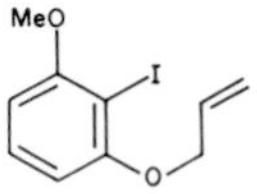	$^nBu_3SnH$, 80°C	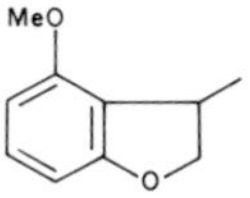	80	[76]
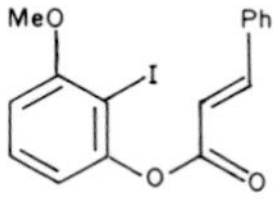	$^nBu_3SnH$, 80°C	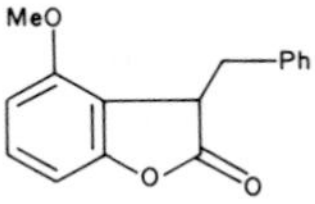	54	[76]
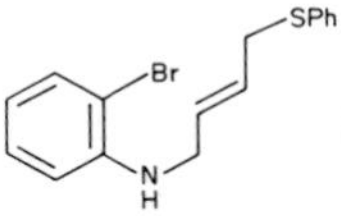	$^nBu_3SnH$, 80°C	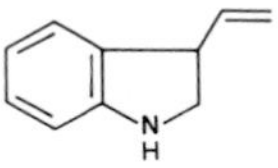	96	[77]
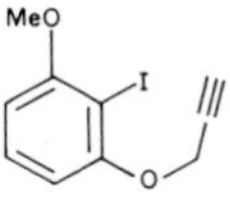	$^nBu_3SnH$, 80°C	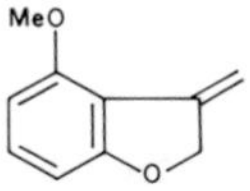	52	[76]
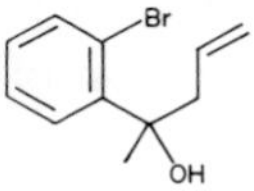	$^nBu_3SnH$, 30°C	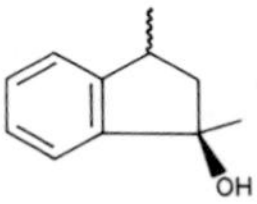	88	[78]
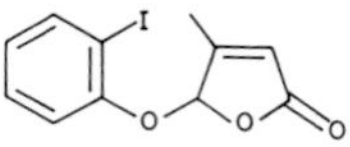	$^nBu_3SnH$, hv	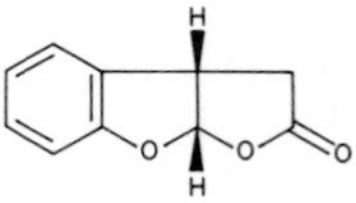	58	[79]

Alternatively, the siting of an electron withdrawing group at the *ortho* position promotes ring expansion of the initial 5-*exo* product [63, 80, 89].

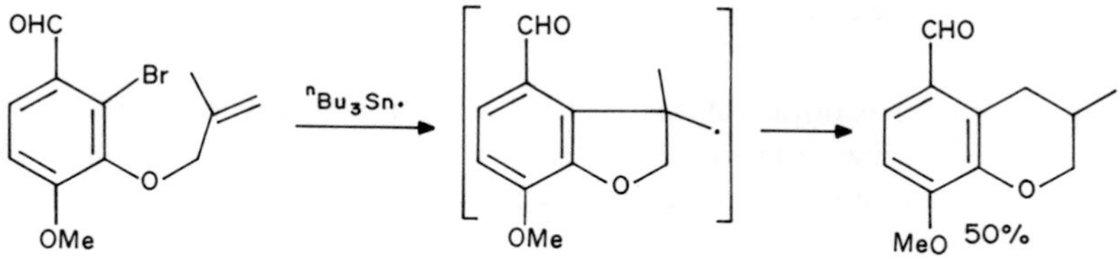

Procedures for aryl radical cyclizations

These are the same as those adopted for the generation and cyclization of 5-hexenyl radicals from 5-hexenyl bromides and iodides (Section 7.1.5).

7.1.9 Alkoxycarbonyl, acyl and related cyclizations

A recent development has involved the cyclization of alkoxycarbonyl and acyl radicals onto alkenes and alkynes in the 5-*exo* mode, permitting the formation of butyrolactones and cyclopentanones, respectively. The optimum method for the generation of these reactive intermediates is by AIBN initiated reaction of tin hydrides with phenyl selenocarbonates and acyl phenylselenides, themselves readily available from chloroformates and acyl chlorides. Examples are given in Table 7.8.

It has also been demonstrated that it is possible to generate thionolactones by the reaction of *O*- and/or *S*-butenyl and butynyl dithiocarbonate esters.

TABLE 7.8

Alkoxycarbonyl and acyl radical cyclizations

Substrate	Conditions	Product	Yield (%)	Ref.
	$^nBu_3SnH$, 80°C		80	[81]
	$^nBu_3SnH$, 80°C	*cis*:*trans* = 2.5:1	92	[81]
	$^nBu_3SnH$, 80°C	$E:Z = 1:12$	80	[82]
	$^nBu_3SnH$, 80°C		69	[83]
	$^nBu_3SnH$, 80°C		86	[83]
	$^nBu_3SnH$ 80°C		81	[83]

[84]

>80%

77%   [85]

General procedure for the preparation of α-methylene-γ-lactones

The appropriate oxirane was treated with a lithium acetylide in the presence of boron trifluoride etherate to give a 3-alkyn-1-ol which was converted to the chloroformate by reaction with phosgene and then to the selenocarbonate with phenylselenol and pyridine. Individual solutions of tri-n-butyltin hydride (0.9 mmol) and AIBN (0.085 mmol), both in benzene (10 ml), were then added simultaneously over 2 h to a solution of the selenocarbonate (0.8 mmol) in benzene (60 ml) at reflux. After evaporation of the solvent the α-benzylidene-γ-lactones were recovered by chromatography on silica.

From ref. [82] with permission.

General preparation of selenoesters [86]

To a stirred solution of the acid (10 mmol) in dichloromethane (20 ml) under a nitrogen atmosphere at room temperature was added a solution of triethylamine (10 mmol) in dichloromethane (10 ml). The mixture was stirred for 10 min, then evaporated under reduced pressure, taken up in dry tetrahydrofuran (25 ml) and added under nitrogen with stirring at room temperature to a preformed solution of tri-n-butylphosphine (20 mmol) and phenylseleneyl chloride (20 mmol) in tetrahydrofuran (30 ml). The reaction mixture was stirred at room temperature for a further 2 h then poured onto ether (300 ml) and water (300 ml). The aqueous layer was further extracted with ether (3 × 100 ml) and the combined organic phases washed with water and saturated sodium chloride solution. After drying and evaporation of the ether, chromatography on silica provided the selenoesters.

General procedure for the cyclization of selenoesters [83]

A solution of the selenoester in benzene (0.007–0.01 M) and AIBN (0.05 equiv.) was heated to reflux and treated with tri-n-butyltin hydride (1.2 equiv.) in benzene dropwise over 2 h and the reaction heated to reflux for a further 0.5–1 h. Purification was by chromatography on silica (or sublimation).

7.1.10 Cyclizations onto carbon–heteroatom multiple bonds

Alkyl radicals cyclize efficiently onto nitriles in the 5-*exo*-dig mode giving, after chain transfer and hydrolysis, cyclopentanones in good yield [87]. Note, however, that the attempted formation of a [3.3.0] bicyclooctane gave a poor yield.

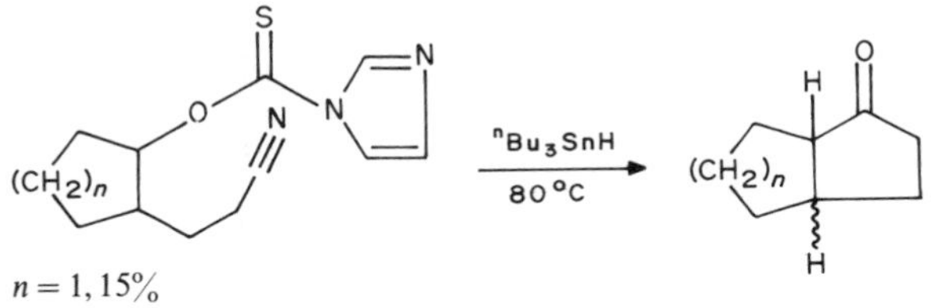

$n = 1, 15\%$
$n = 2, 67\%$
$n = 3, 65\%$

Cyclization onto aldehydes resulting, after chain transfer, in the formation of cyclopentanols is an efficient reaction under the appropriate conditions [88]. This apparently contrathermodynamic process relies on the rapid abstraction of hydrogen from the stannane by the intermediate alkoxy radical [89].

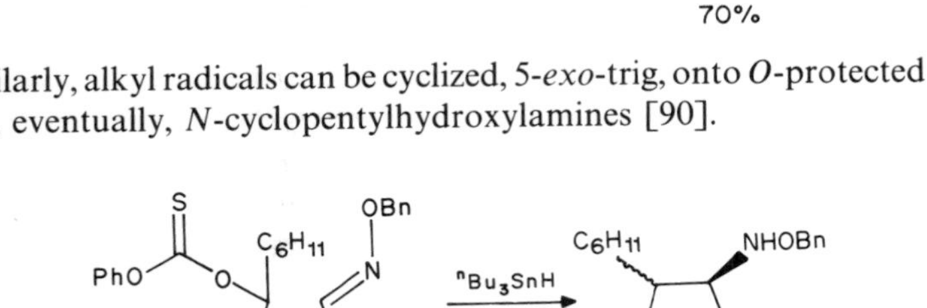

70%

Similarly, alkyl radicals can be cyclized, 5-*exo*-trig, onto *O*-protected oximes giving, eventually, *N*-cyclopentylhydroxylamines [90].

84%, ~1:1

7.2 PREPARATION OF SIX MEMBERED RINGS BY
6-*EXO* CYCLIZATIONS

The formation of six membered rings by free radical chain reactions involving 6-*exo* cyclization of 6-heptenyl and related radicals presents several problems.

Firstly, the rate constant for 6-heptenyl cyclization at 25°C is $5.4 \times 10^3 \, s^{-1}$, some 40 times slower than that of the 5-hexenyl cyclization [1]. The obvious consequence being that chain transfer by attack of the ring open radical on, for example, the stannane is a much more effective competing process than with the 5-hexenyl radical.

The second potential problem (which has been translated into a useful synthetic process, see Section 7.6) is 1,5-hydrogen abstraction leading to a resonance stabilized allyl radical.

Furthermore, *endo* mode cyclization of the simple 6-heptenyl radical is only approximately seven times less rapid than *exo* mode cyclization at 25°C, and it is to be expected, therefore, that the formation of *endo* cyclization products is a more serious competing reaction than in the 5-hexenyl case.

In order to overcome these difficulties the rate of *exo* mode cyclization must be increased. This has commonly been achieved by one of two methods: (i) restriction of degrees of freedom in the heptenyl chain either by the incorporation of rings or by judicious use of the "Thorpe–Ingold" effect or (ii) activation of the double bond. Recently, an increasing number of aryl radical cyclizations have been described which provide six membered rings in high yield via the 6-heptenyl type cyclization. This is readily understood in terms of the high rate constant for cyclization of such systems [1].

$$3 \times 10^8 s^{-1}$$

All of the examples in Table 7.9 benefit from one or the other, if not all, of these rate enhancing factors.

TABLE 7.9

6-Heptenyl cyclizations

Substrate	Conditions	Product	Yield (%)	Ref.
	^{n}Bu$_3$SnH, hexane	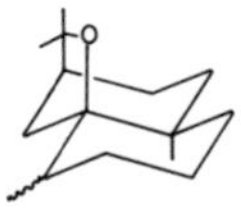	67	[91]
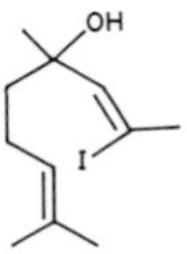	^{n}Bu$_3$SnH, 80°C	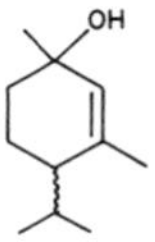	72	[61]
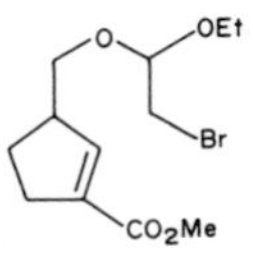	^{n}Bu$_3$SnH, 80°C	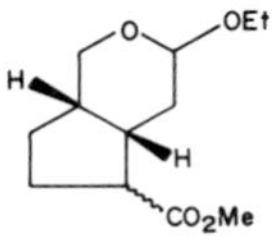 endo:exo = 5.5:1	90	[38]
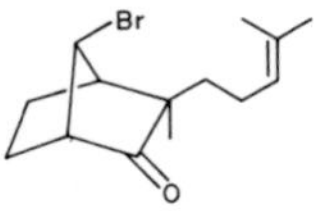	^{n}Bu$_3$SnH, 36°C	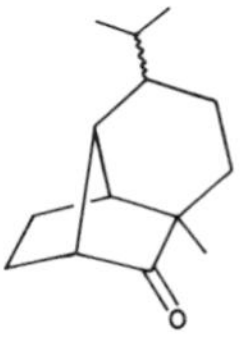	62	[92]
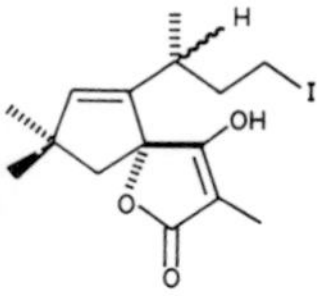	^{n}Bu$_3$SnH 80°C	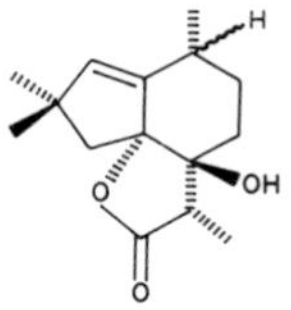	33	[93]
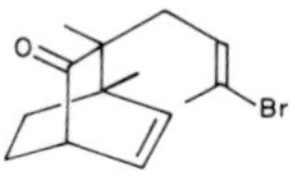	^{n}Bu$_3$SnH, 80°C	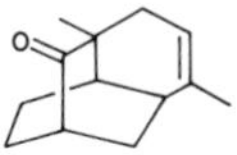	70	[94]
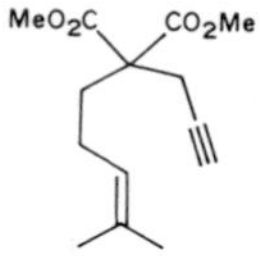	(i) ^{n}Bu$_3$SnH, 80°C (ii) SiO$_2$	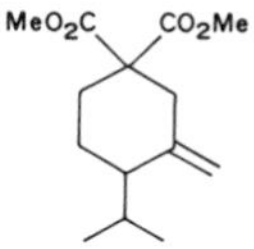	55	[71]

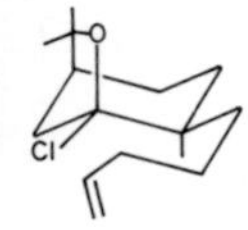

TABLE 7.9 *Continued*

TABLE 7.9 *Continued*

Substrate	Conditions	Product	Yield (%)	Ref.
	$^nBu_3SnH$, 80°C	$Z:E = 14:1$	35	[95]
	(i) $Hg(OAc)_2$, HOAc (ii) $NaBH_4$ (iii) Jones		44	[18]
	(i) $Hg(OAc)_2$, HOAc (ii) $NaBH_4$		10	[18]
	(i) $Hg(OAc)_2$, HOAc (ii) $NaBH_4$		77	[18]
	$^nBu_3SnH$, 80°C		68	[96]
	$^nBu_3SnH$, 80°C (Benzene-refluxing temp.)		75	[97]
	$^nBu_3SnH$, 80°C	$exo:endo = 7:1$	60	[43]
	$^nBu_3SnH$, Δ, hexane		72	[91]

TABLE 7.9 *Continued*

Substrate	Conditions	Product	Yield (%)	Ref.
	Ph$_3$SnH Δ, hexane	~1:1	79	[87]
	^{n}Bu$_3$SnH, 80°C	~ 1:1	90	[98]
	^{n}Bu$_3$SnH, 80°C	~ 1:1	84	[83]
	^{n}Bu$_3$SnH, 80°C		83	[83]
	^{n}Bu$_3$SnH, 110°C PhSnH, 80°C (C$_6$H$_6$-reflux)	Products 37 , 55		[99]
	^{n}Bu$_3$SnH, 80°C		70	[100]
	^{n}Bu$_3$SnH, 80°C		62	[101]
	Δ, THF ^{t}BuSH *hv*		60	[102]

Procedures for 6-heptenyl cyclizations

These are the same as those for 5-hexenyl cyclizations (Section 7.1.5).

A noteworthy development is the realization that six membered rings may be formed efficiently by *exo* mode cyclization onto aldehydes by the tin hydride method [88, 103].

This mode of cyclization has been demonstrated to be more rapid than the 5-hexenyl type cyclization [103].

Formation of cyclohexanols from 6-iodohexanals [103]

In a typical experiment the radicals were generated by treating a 0.03 M solution of the iodide in benzene with 1 equiv. of tri-n-butyltin hydride and a catalytic amount of AIBN under reflux in an argon atmosphere.

7.3 PREPARATION OF SEVEN MEMBERED AND MEDIUM SIZED RINGS

7.3.1 Cyclizations

The problems associated with the formation of medium sized rings by ring closure reactions are widely appreciated. Free radical cyclizations are no exception. Even the formation of seven membered rings in the 7-*exo*-trig mode is likely to be unsuccessful in most cases as the rate constant at 25°C for 8-octenyl cyclization is $< 7 \times 10^{-1} s^{-1}$ [1]. Nevertheless, Boger was able to prepare a seven membered ring by acyl radical cyclization with a substrate that incorporated only a limited number of degrees of freedom and which carried an activated alkene [83].

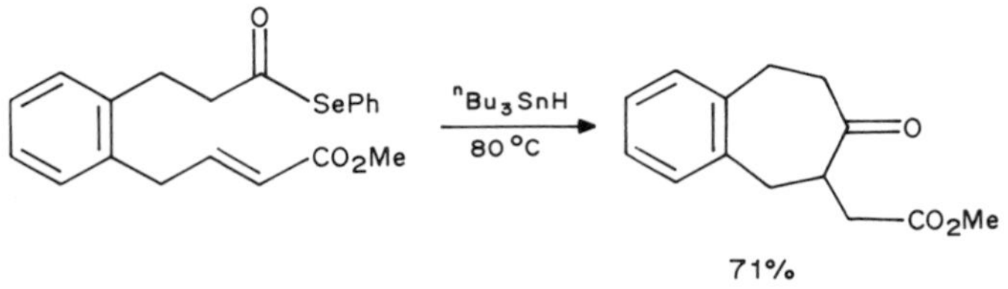

71%

7.3.2 Ring expansions

Radical ring expansion methods allow relatively facile entry into seven membered and medium ring compounds. The general principle involving one carbon ring expansion is illustrated, where the group Y is either a methylene group (as in overall *endo* cyclization of vinyl radicals, Section 7.1.6) or an oxygen atom and the group Z is some group capable of stabilizing the ring expanded radical.

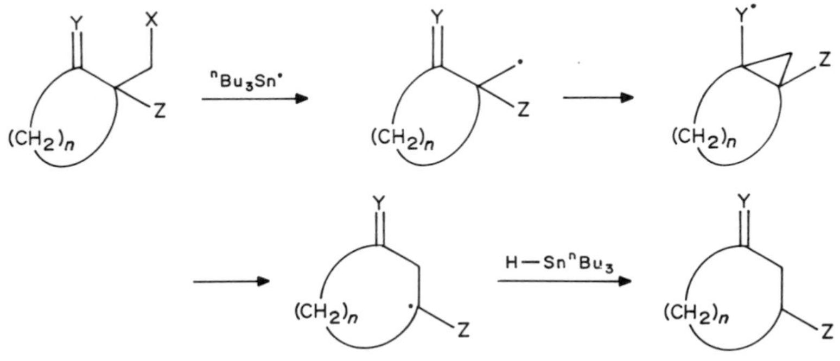

Ring expansions of this kind were reported in 1961 [104] (five to six membered ring expansion) and again in 1984 [105]. Systematic investigations have more recently been carried out by Beckwith [63] and Dowd [106].

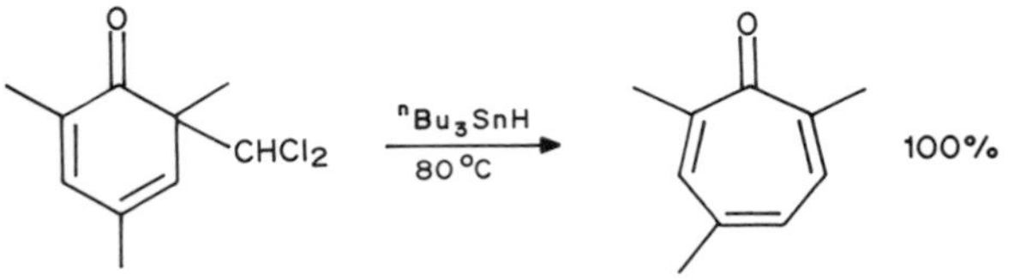

100%   [105]

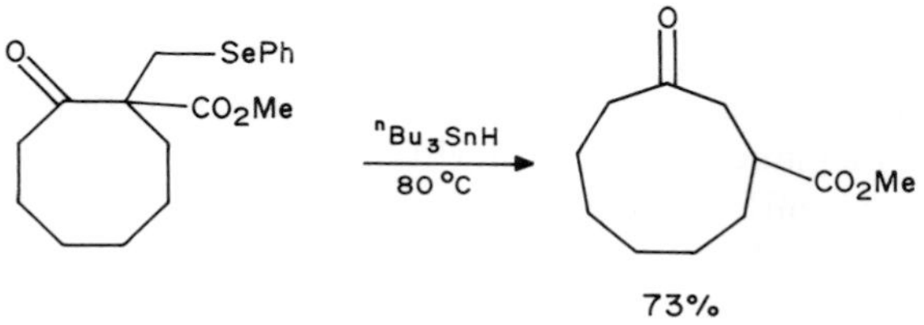

[63]

73%

General procedure for one carbon ring expansion:
3-carbomethoxycyclononanone [63]

Methyl 2-oxocyclooctane-1-carboxylate (920 mg, 5 mmol) was taken up in methanol (3 ml) and a 2 M solution of sodium methoxide in methanol (2.6 ml, 5.2 mmol) was added dropwise. The solution was stirred at room temperature for 10 min. Chloro(phenylseleno)methane (1.13 g, 5.5 mol) and a catalytic amount of sodium iodide were added and the solution heated under reflux for 3 h. The bulk of the solvent was removed under reduced pressure and the residue in ether (10 ml) washed with 10% aqueous hydrochloric acid. Concentration and chromatography gave 59% of the selenomethyl derivative.

The selenide (1 mmol) was taken up in benzene (10 ml) and the solution purged with nitrogen for 10 min and then brought to reflux. A solution of tri-n-butyltin hydride (1.1 mmol) and AIBN (0.1 mmol) in benzene (5 ml) was then added over 3 h by means of a syringe pump. After a further 30 min at reflux the benzene was removed and the residues in ether (10 ml) treated with a 10% aqueous solution of potassium fluoride (5 ml) for 10 min. The organic layer was filtered, dried, evaporated and chromatographed on silica (20% ethyl acetate in hexane) giving the product as a clear oil.

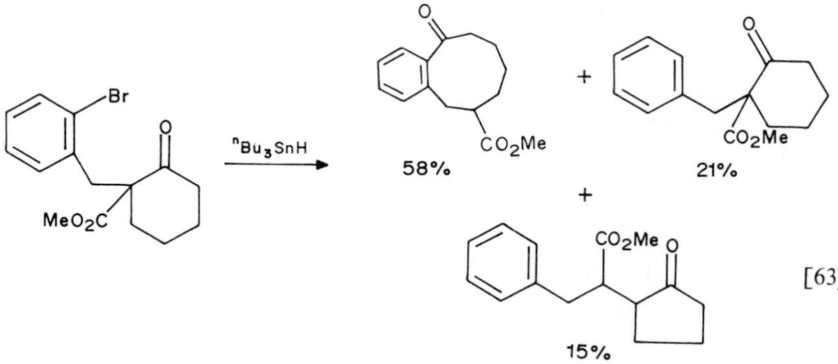

58%   21%

15%

[63]

The method also allows ring expansion by three and four carbons [63, 106]. The electron withdrawing group stabilizing the ring expanded radical is essential if high yields are to be obtained [107].

The problem of the reversible nature of the key bond fission in these reactions has been overcome by Baldwin by incorporation of a stannyl group β to the ring expanded radical such that stannyl radical elimination occurs and drives the equilibrium in the desired direction [108].

The formation of *trans*-alkenes from *trans* substituted precursors and *cis*-alkenes from *cis* precursors can be rationalized on stereoelectronic grounds.

7.4   PREPARATION OF LARGE RINGS

The formation of 10–20 membered rings by radical cyclizations has been demonstrated by Porter [109]. Cyclizations of this kind, which take place in the *endo* mode, are highly comparable to the intermolecular addition of radicals to alkenes (Chapter 6), thus it is essential that the alkene be activated and terminal if a good yield is to be obtained, as in the examples shown in Table 7.10.

Radical macrocyclization [111]

The most convenient conditions for radical macrocyclization involved heating of the iodide (3–6 mM) in dry benzene, along with 1.1 equiv. of tri-n-butyltin hydride and 0.1 equiv. of AIBN, to reflux under argon for 3 h. Solvent removal and chromatography on silica provided the cyclization and reduction products.

TABLE 7.10

Radical macrocyclizations

Substrate	Conditions	Product	Yield (%)	Ref.
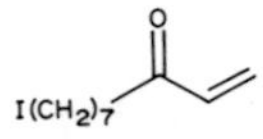 $I(CH_2)_7$ (acryloyl ketone)	$^nBu_3SnH$, 80°C	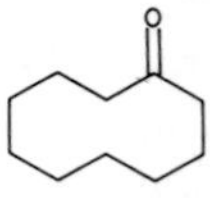 cyclic ketone	15	[109]
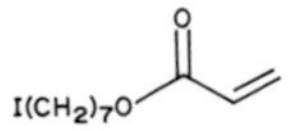 $I(CH_2)_7O$ (acrylate ester)	$^nBu_3SnH$, 80°C	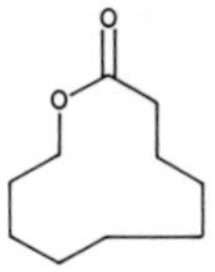 macrolactone	15–20	[109]
$I(CH_2)_{11}COCH{=}CH_2$	$^nBu_3SnH$, 80°C	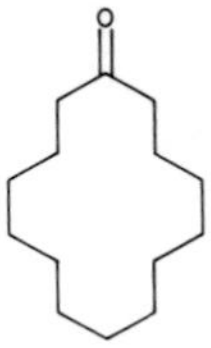 cyclic ketone	65	[109]
$I(CH_2)_{15}COCH{=}CH_2$	$^nBu_3SnH$, 80°C	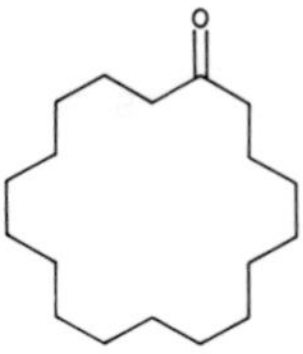 cyclic ketone	54	[109]
$I(CH_2)_7C{\equiv}C(CH_2)_2COCH{=}CH_2$		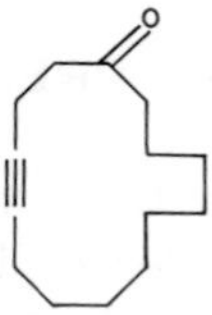 cyclic alkyne ketone	76	[109]
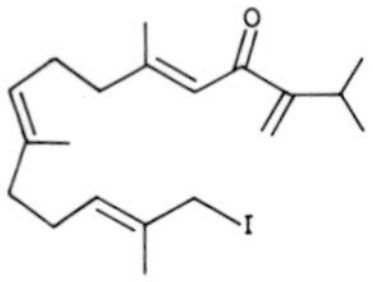 (triene iodide structure)	$^nBu_3SnH$, 80°C	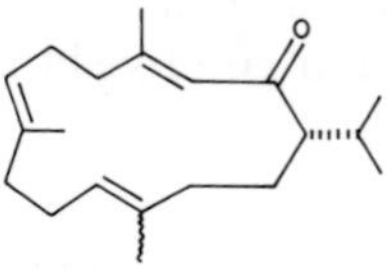 cyclic product, $E{:}Z = 4{:}1$	40	[110]

7.5 TRANSANNULAR CYCLIZATIONS

In some instances transannular radical cyclizations provide a useful entry into fused bicyclic systems. The smallest system that may be constructed in this manner is the bicyclo[3.3.0]octane system and, although it has been shown [112] that under typical preparative conditions the parent cyclooct-4-enyl radical is reluctant to cyclize, substituted cyclooct-4-enyl radicals do so relatively easily. Examples of the formation of bicyclic systems by transannular cyclizations are given in Table 7.11.

TABLE 7.11

Transannular cyclizations

Substrate	Conditions	Product	Yield (%)	Ref.
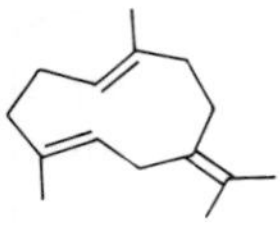	PhSH, hv	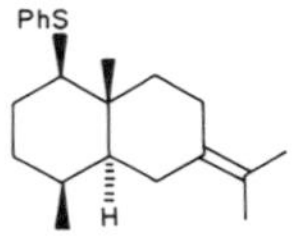	34	[113]
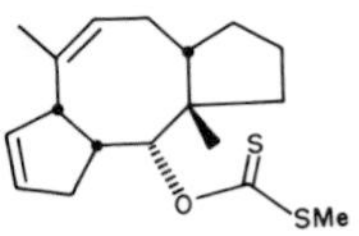	^{n}Bu$_3$SnH, 110°C	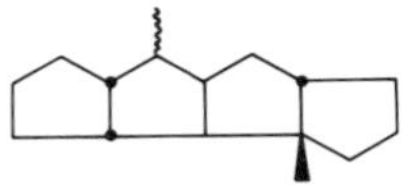	67	[114]
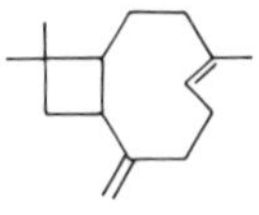	CH$_3$CHO 125°C	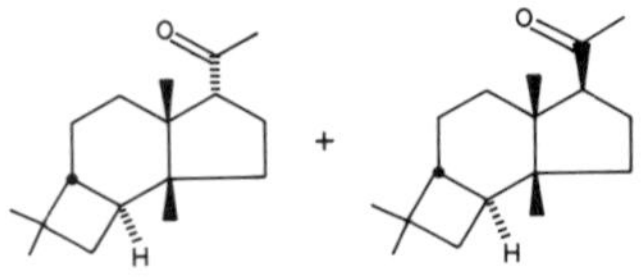	30, 6	[115]
		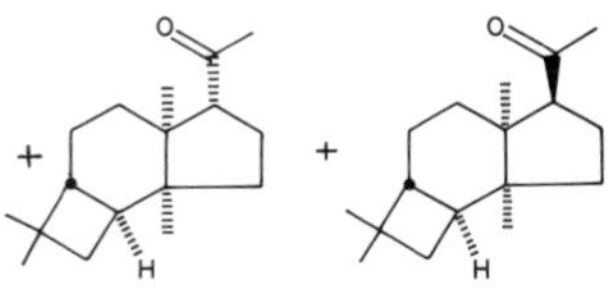	12, 6	
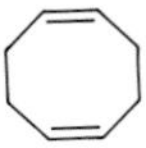	(EtO)$_2$PH, (tBuO)$_2$	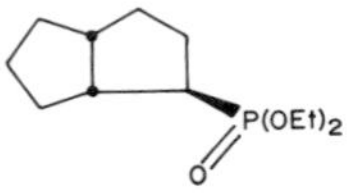	60	[116]
	TsCl, hv	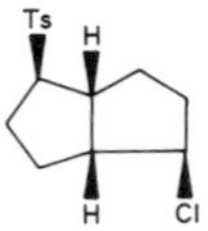	44	[117]

Preparation of 2-chloro-6-p-toluenesulphonylbicyclo[3.3.0]octane from cyclooocta-1,5-diene [117]

Into a Pyrex reaction vessel were introduced cycloocta-1,5-diene (5 mmol), *p*-toluenesulphonyl chloride (5 mmol), acetonitrile (140 ml) and 10 mol% of AIBN. The externally cooled reaction mixture was then maintained under an inert atmosphere and irradiated from the outside with a Hanau Q81 mercury lamp for 15 h. The solvent was then evaporated and the residue purified by chromatography on silica to give the title compound (55%), m.p. 106°C.

Further examples of transannular cyclizations are given in Section 7.8.

7.6 HYDROGEN ATOM ABSTRACTION/CYCLIZATION

A recent addition to the synthetic chemists armoury is the hydrogen atom abstraction/cyclization process dubbed "translocation of radical sites by 1,5-hydrogen abstraction". In this useful methodology an electrophilic radical generated, for example from an aryl or vinyl halide, abstracts a hydrogen atom from the δ-position, providing a new radical which then cyclizes onto an appropriately placed doubled bond as illustrated in Parsons' synthesis of a pyrrolizidine precursor [118].

60—85%

Further examples were provided by Curran. From a synthetic point of view the most useful is perhaps the use of *o*-bromophenylsilyl ethers as combined radical sources/protecting groups enabling the generation α-alkoxy radicals [119].

61%
(cis : trans, 1 : 1.1)

Ethyl (2-dimethylphenylsiloxycyclopentyl)acetate from ethyl
7-hydroxyhept-2E-enoate by radical translocation/cyclization [119]

Ethyl 7-hydroxyhept-2*E*-enoate is silylated with (2-bromophenyl)dimethyl-
silyl chloride and the product reacted with 10 mol % of tri-n-butyltin chloride,
10 mol % of AIBN and 2 equiv. of sodium cyanoborohydride in t-butyl alcohol
(0.05 M) at reflux for 4–6 h

7.7 FORMATION OF *ENDO* MODE PRODUCTS BY CYCLIZATION OF STABILIZED RADICALS

Pioneering work in the radical cyclization field was carried out by Julia with
stabilized radicals [120]. In common with other radical cyclizations, these
reactions proceed kinetically in the *exo* mode, but at moderately high
temperatures and in solvents of poor hydrogen donor capacity the cyclizations
are reversible and the thermodynamic *endo* mode products are isolated
(Table 7.12).

TABLE 7.12

Effect of temperature on malonomononitrile cyclizations

	Yield (%)	
Conditions	**A**	**B**
$-70°C$ (hv, [acetone])	50	50
25°C	35	65
105°C	29	71

Reactions of this kind can therefore provide a ready route to six and even
seven membered rings [120].

Preparation of ethyl 1-cyano-trans-decalin-1-carboxylate [121, 122]

Ethyl 1-cyano-5-cyclohex-1-enyl-pentanoate (12 g) in cyclohexane (500 ml) was added to cyclohexane (500 ml) at reflux, to which was added benzoyl peroxide (3 g) in four aliquots over 8 h. The reaction was heated to reflux overnight and then for a further day with further additions of benzoyl peroxide (3 g). Distillation yielded 6.8 g of a product which was treated with potassium permanganate solution until a persistent colour was obtained. Extraction and redistillation provided the cyanoester (4.8 g, 40%), b.p. 110–113°C/63 mmHg.

7.8 TANDEM PROCESSES

So-called tandem processes involving multiple radical cyclizations provide a rapid entry into sometimes complex molecular frameworks. An elegant example is provided by the entry of Stork into the C–D ring of steroidal butenolides [123].

The power of such tandem radical cyclizations is especially well demonstrated by the syntheses, by the Curran group, of both linear and angular triquinanes [124] and by the Parsons approach to the southern part of the milbemycin system [125].

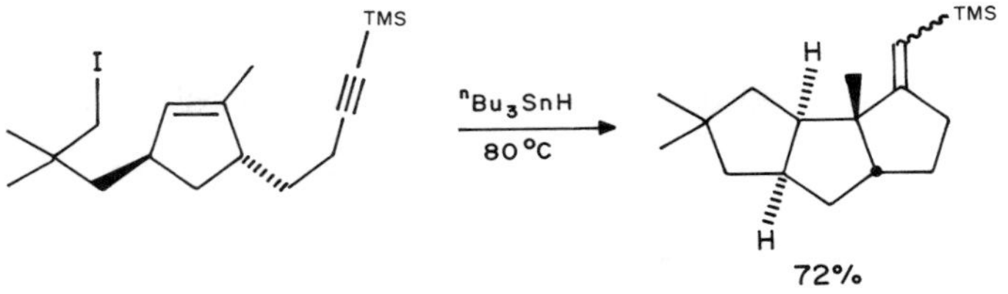

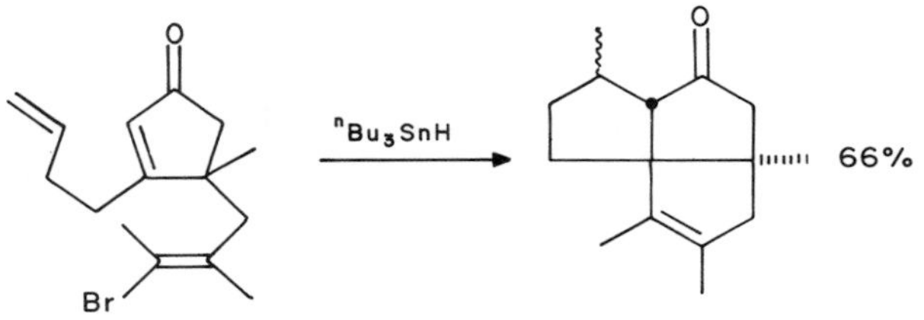

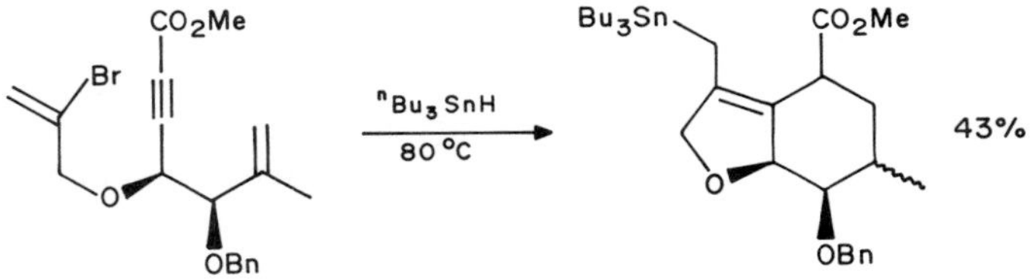

72%

66%

43%

More complex systems have been investigated by Beckwith [126].

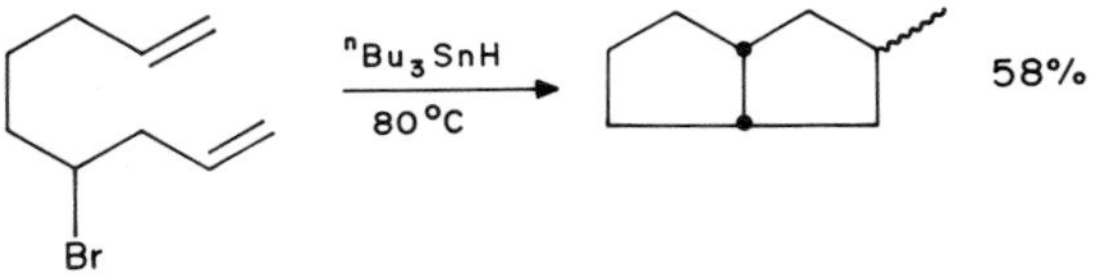

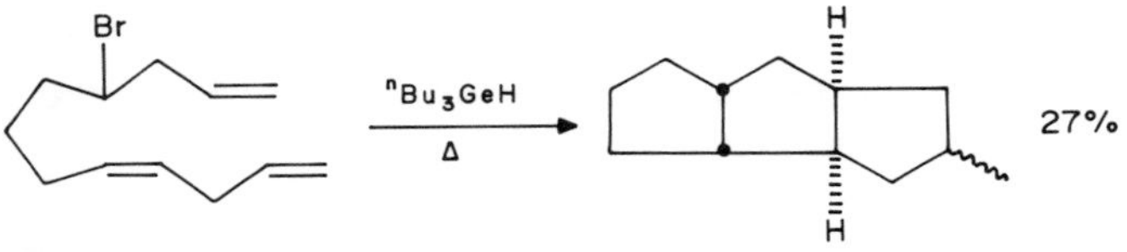

58%

27%

The Julia type cyclizations of stabilized radicals may also be applied in tandem processes [127].

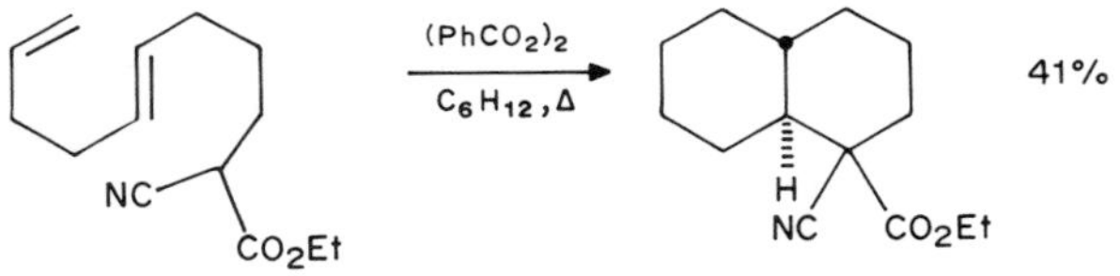

41%

Indeed, reversibility in the first cyclization was found to be essential in the Winkler entry into linear triquinanes, as the more usual tin hydride method gave only low yields of the transannularly cyclized product [128].

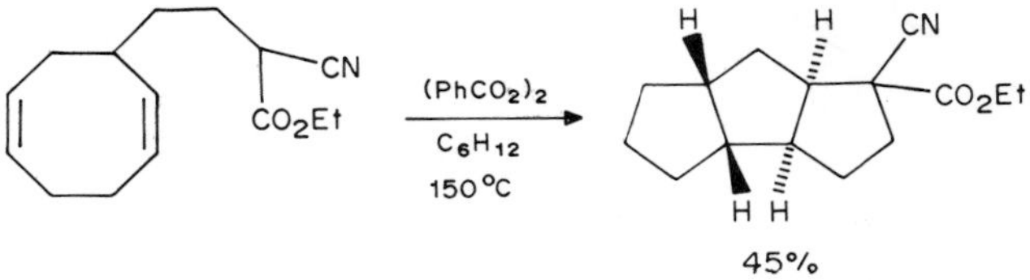

45%

Tandem radical processes have also been devised which incorporate both inter- and intramolecular radical carbon–carbon bond forming steps [84].

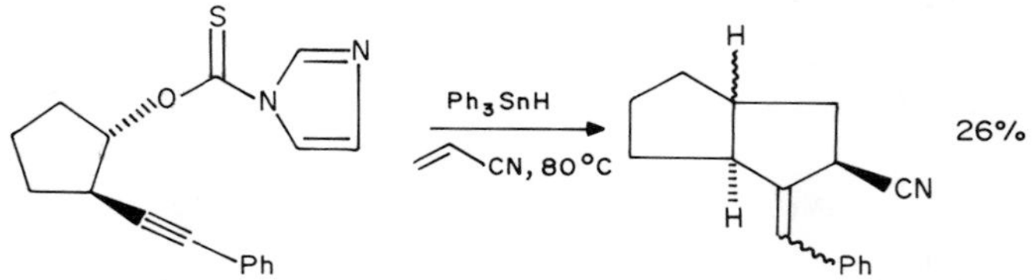

26%

This concept has been extended by Barton to include two inter- and one intramolecular carbon–carbon bond forming steps, enabling the rapid assembly of bicyclic systems from very simple precursors [129].

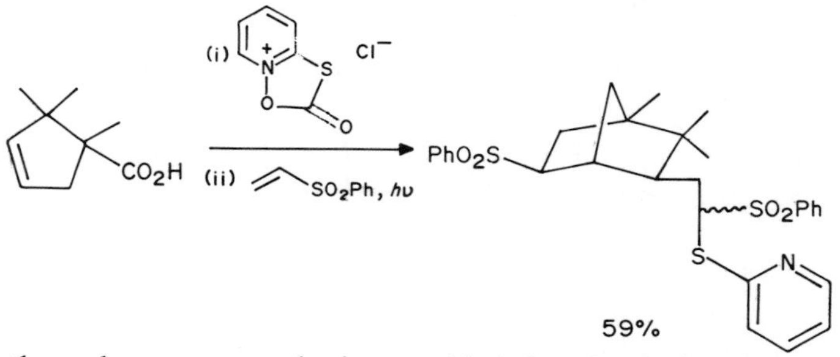

59%

A further elegant example is provided by Stork in the context of prostaglandin synthesis [130].

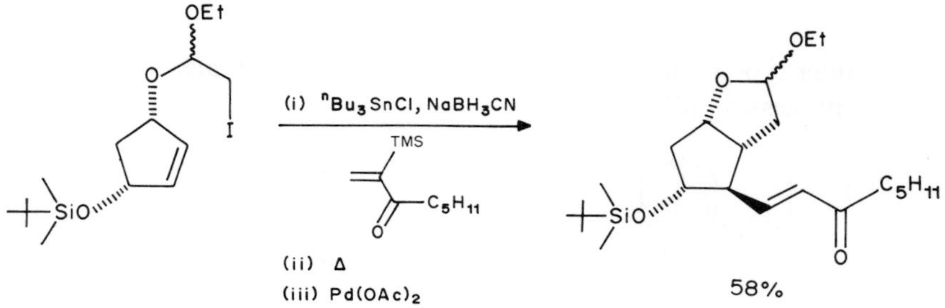

58%

Finally, several authors have described the coupling of cyclopropylmethyl ring opening sequences with 5-hexenyl cyclizations.

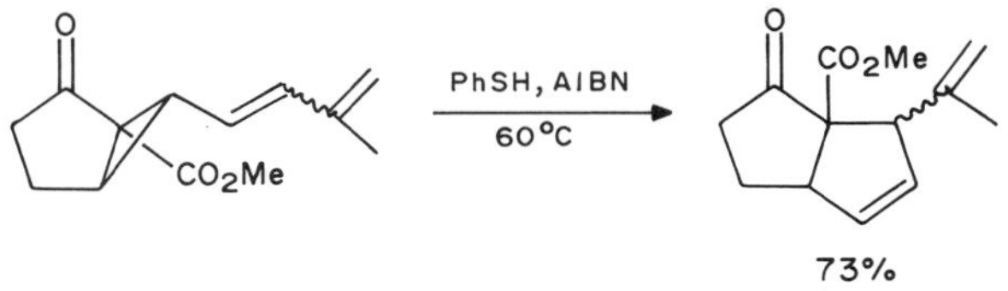

[131]

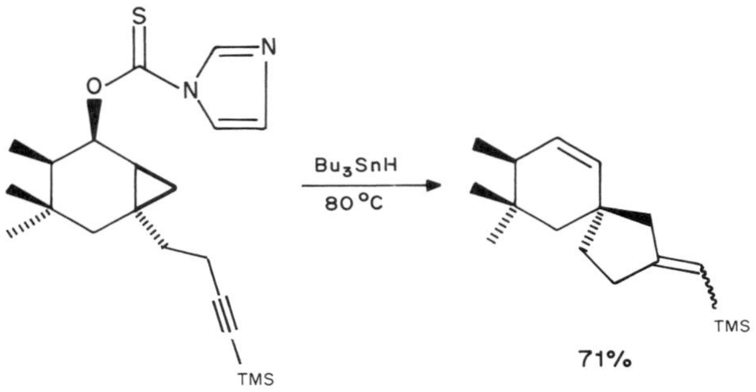

73%

[142]

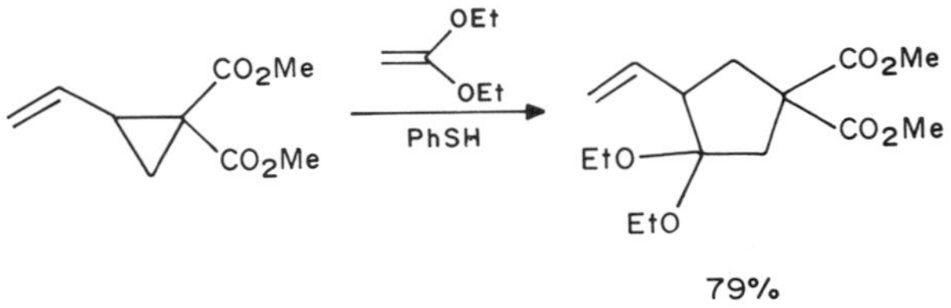

71%

[133]

79%

Simple tandem procedures coupling two 5-hexenyl cyclizations in sequence are experimentally straightforward and standard protocols for 5-hexenyl cyclizations (Section 7.1.5) should be adopted. Sequences coupling inter- with intramolecular C–C bond forming steps are best optimized individually.

7.9   RADICAL CYCLIZATIONS BY S_H2 AT CARBON

Johnson has developed some very useful, though little appreciated chemistry, for the formation of three and five membered rings involving attack of electrophilic radicals at multiple bonds followed by homolytic substitution at carbon with loss of the cobaloxime radical, which subsequently acts as a chain transfer agent. The precursor cobaloximes are air stable orange solids which are simple to prepare. The reactions are carried out simply by heating in dichloromethane. Representative examples are given in Table 7.13 [134].

TABLE 7.13

Cyclizations by S_H2 at carbon

Substrate	Conditions	Product	Yield (%)
[Co]	Br_2CHCN, 60°C	NC, Br	61
[Co]	$ArSO_2I$, CH_2Cl_2, Δ	ArSO₂ (1:1)	80
[Co]	$ArSO_2I$, CH_2Cl_2, Δ	ArSO₂	80
[Co]	CCl_4, 80°C	CCl₃	

$[Co] = Co(\text{III})(dmgH)_2(py).$

REFERENCES

1. A. L. J. Beckwith and C. H. Schiesser, *Tetrahedron* **41**, 3925 (1985); A. L. J. Beckwith, *Tetrahedron* **37**, 3073 (1981) and references therein.
2. A. L. J. Beckwith, T. Lawrence and A. K. Serelis, *J. Chem. Soc. Chem. Commun.* 484 (1980).
3. A. L. J. Beckwith, I. Blair and G. Phillipou, *J. Am. Chem. Soc.* **96**, 1613 (1974).
4. S. Wolff and W. C. Agosta, *J. Chem. Res. (S)* 78 (1981).
5. M. A. M. Bradney, A. D. Forbes and J. Wood, *J. Chem. Soc. Perkin Trans. 2* 1655 (1973).
6. T. V. RajanBabu, T. Fukunaga and G. S. Reddy, *J. Am. Chem. Soc.* **111**, 1759 (1989) and references therein.
7. N. O. Brace, *J. Org. Chem.* **34**, 2441 (1969) and references therein.
8. C. S. Wilcox and L. M. Thomasco, *J. Org. Chem.* **50**, 546 (1985).
9. C. Descoins, M. Julia and H. V. Sang, *Bull. Soc. Chim. Fr.* 4087 (1971).
10. R. Tsang and B. Fraser-Reid, *J. Am. Chem. Soc.* **108**, 2116, 8102 (1986).
11. D. L. J. Clive, D. R. Cheshire and L. Set, *J. Chem. Soc. Chem. Commun.* 353 (1957).
12. D. J. Hart and H. C. Huang, *Tetrahedron Lett.* **26**, 3749 (1985); C. P. Chuang, J. C. Gallucci and D. J. Hart, *J. Org. Chem.* **53**, 3210 (1988); C. P. Chuang, J. C. Gallucci, D. J. Hart and C. Hoffman, *J. Org. Chem.* **53**, 3218 (1988).
13. H. Hashimoto, F. Furuichi and T. Miwa, *J. Chem. Soc. Chem. Commun.* 1002 (1987).
14. D. J. Hart and Y. M. Tsai, *J. Org. Chem.* **47**, 4403 (1982).
15. F. E. Ziegler and Z. Zheng, *Tetrahedron Lett.* **28**, 5973 (1987).
16. T. L. Fevig, R. L. Elliott and D. P. Curran, *J. Am. Chem. Soc.* **110**, 5064 (1988); V. K. Yadav and A. G. Fallis, *Tetrahedron Lett.* **29**, 897 (1988).
17. G. Stork and M. E. Reynolds, *J. Am. Chem. Soc.* **110**, 6911 (1988).

18. S. Danishefsky, S. Chackalamannil and B. J. Uang, *J. Org. Chem.* **47**, 2231 (1982).
19. D. H. R. Barton, D. Crich and G. Kretzschmar, *J. Chem. Soc. Perkin Trans. 1*, 39 (1986).
20. G. A. Kraus and K. Landgrebe, *Tetrahedron* **41**, 4039 (1985).
21. D. P. Curran and C. T. Chang, *J. Org. Chem.* **54**, 3140 (1989).
22. N. O. Brace, *J. Org. Chem.* **31**, 2879 (1966).
23. Y. Ueno, C. Tanaka and M. O. Okawara, *Chem. Lett.* 795 (1983).
24. A. Padwa, H. Nimmesgern and G. S. K. Wong, *Tetrahedron Lett.* **26**, 957 (1985).
25. N. Ono, H. Miyake, A. Kamimura, I. Hamamoto, R. Tamura and A. Kaji, *Tetrahedron* **41**, 4013 (1985).
26. E. Castagnino, S. Corsano and D. H. R. Barton, *Tetrahedron Lett.* **30**, 2983 (1989).
27. Y. Ueno, R. K. Khare and M. Okawara, *J. Chem. Soc. Perkin Trans. 1*, 2637 (1983).
28. D. S. Middleton, N. S. Simpkins, M. J. Begley and N. K. Terrett, *Tetrahedron* **46**, 545 (1990).
29. S. J. Danishefsky and J. Panek, *J. Am. Chem. Soc.* **109**, 917 (1987).
30. T. Harrison, G. Pattenden and P. L. Myers, *Tetrahedron Lett.* **29**, 3869 (1988).
31. Y. Watanake, Y. Ueno, C. Tanaka, M. Okawara and T. Endo, *Tetrahedron Lett.* **28**, 3953 (1987).
32. T. A. K. Smith and G. H. Whitham, *J. Chem. Soc. Perkin Trans. 1*, 313, 319 (1989).
33. A. L. J. Beckwith and P. E. Pigou, *J. Chem. Soc. Chem. Commun.* 85 (1986).
34. S. Danishefsky and E. Taniyama, *Tetrahedron Lett.* **24**, 15 (1983).
35. D. L. J. Clive and P. L. Beaulieu, *J. Chem. Soc. Chem. Commun.* 307 (1983).
36. R. S. Jolly and T. Livinghouse, *J. Am. Chem. Soc.* **110**, 7536 (1988).
37. J. L. Belletire and N. O. Mahmoodi, *Tetrahedron Lett.* **30**, 4363 (1989).
38. G. Stork, R. Mook, S. A. Biller and S. D. Rychnovsky, *J. Am. Chem. Soc.* **105**, 3741 (1983); Y. Ueno, K. Chino, M. Watanabe, O. Moriya and M. Okawara, *J. Am. Chem. Soc.* **104**, 5565 (1982).
39 M. J. Begley, H. Bhandal, J. H. Hutchinson and G. Pattenden, *Tetrahedron Lett.* **28**, 1317 (1987).
40. G. Stork and R. Mah, *Tetrahedron Lett.* **30**, 3609 (1989).
41. G. Stork and R. Mah, *Heterocycles* **28**, 723 (1989).
42. A. L. J. Beckwith and D. R. Boate, *Tetrahedron Lett.* **26**, 1761 (1985).
43. D. A. Burnet, J. K. Choi, D. J. Hart and Y. M. Tsai, *J. Am. Chem. Soc.* **106**, 8201 (1984); D. J. Hart and Y. M. Tsai, *J. Am. Chem. Soc.* **106**, 8209 (1984); see also S. Kano, Y. Yuasa, K. Asami and S. Shibuya, *Heterocycles* **27**, 1437 (1988).
44. M. D. Bachi, A. De Mesmaeker and N. S. De Mesmaeker, *Tetrahedron Lett.* **28**, 2637, 2887 (1987).
45. G. E. Keck and E. J. Enholm, *Tetrahedron Lett.* **26**, 3311 (1985).
46. H. Nishiyama, T. Kitajima, M. Matsumoto and K. Itoh, *J. Org. Chem.* **49**, 2298 (1984).
47. G. Stork and M. Kahn, *J. Am. Chem. Soc.* **107**, 500 (1985); M. T. Crimmins and R. O'Mahony, *J. Org. Chem.* **54**, 1157 (1989).
48. M. Koreeda and I. A. George, *J. Am. Chem. Soc.* **108**, 8098 (1986).
49. J. W. Wilt, *Tetrahedron* **41**, 3979 (1985).
50. J. K. Crandall and D. J. Keyton, *Tetrahedron Lett.* 1653 (1969).
51. M. D. Bachi and E. Bosch, *Tetrahedron Lett.* **27**, 641 (1986).
52. O. Moriya, M. Okawara and Y. Ueno, *Chem. Lett.* 1437 (1984).
53. J.-K. Choi and D. J. Hart, *Tetrahedron* **41**, 3959 (1985).
54. L. Set, D. R. Cheshire and D. L. J. Clive, *J. Chem. Soc. Chem. Commun.* 1205 (1985).
55. S. M. Bennett and D. L. J. Clive, *J. Chem. Soc. Chem. Commun.* 878 (1986).
56. J. P. Dulcere, J. Rodriguez, M. Santelli and J. P. Zahra, *Tetrahedron Lett.* **28**, 2009 (1987).
57. E. J. Corey and M. M. Mehrotra, *Tetrahedron Lett.* **29**, 57 (1988).
58. G. Haaima and R. T. Weavers, *Tetrahedron Lett.* **29**, 1085 (1988).

59. J. M. Clough, G. Pattenden and P. G. Wight, *Tetrahedron Lett.*, **30**, 7469 (1989).
60. M. Apparu and J. K. Crandall, *J. Org. Chem.* **49**, 2125 (1984).
61. G. Stork and N. H. Baine, *J. Am. Chem. Soc.* **104**, 2321 (1982).
62. A. L. J. Beckwith and D. M. O'Shea, *Tetrahedron Lett.* **27**, 4525 (1986); G. Stork and R. Mook, *Tetrahedron Lett.* **27**, 4529 (1986).
63. A. L. J. Beckwith, D. M. O'Shea and S. W. Westood, *J. Am. Chem. Soc.* **110**, 2565 (1988).
64. O. Moriyama, M. Okawara and Y. Ueno, *Chem. Lett.* 1437 (1984).
65. H. Urabe and I. Kuwajima, *Tetrahedron Lett.* **27**, 1355 (1986).
66. N. N. Marinovic and H. Ramanathan, *Tetrahedron Lett.* **24**, 1871 (1983).
67. D. Crich and S. M. Fortt, *Tetrahedron Lett.* **28**, 2895 (1987).
68. J. Knight and P. J. Parsons, *J. Chem. Soc. Perkins Trans. 1*, 1237 (1987).
69. A. P. Neary and P. J. Parsons, *J. Chem. Soc. Chem. Commun.* 1090 (1989).
70. S. P. Munt and E. J. Thomas, *J. Chem. Soc. Chem. Commun.* 480 (1989).
71. G. Stork and R. Mook, *J. Am. Chem. Soc.* **109**, 2829 (1987).
72. R. Mook and P. M. Sher, *Organic Synthesis* **66**, 75 (1987).
73. K. Nozaki, K. Oshima and K. Utimoto, *J. Am. Chem. Soc.* **109**, 2547 (1987).
74. J. Ardisson, J. P. Férézou, M. Julia, L. Lenglet and A. Pancrazi, *Tetrahedron Lett.* **28**, 1997 (1987).
75. E. Lee, S. B. Ko, K. W. Jung and M. H. Chang, *Tetrahedron Lett.* **30**, 827 (1989).
76. K. Shankaran, C. P. Sloan and V. Snieckus, *Tetrahedron Lett.* **26**, 6001 (1985).
77. Y. Ueno, K. Chino and M. Okawara, *Tetrahedron Lett.* **23**, 2575 (1982); D. L. Boger and R. J. Wysocki, *J. Org. Chem.* **54**, 1238 (1989).
78. J. K. Macleod and L. C. Monahan, *Tetrahedron Lett.* **29**, 391 (1988).
79. S. Wolff and H. M. R. Hoffmann, *Synthesis* 760 (1988); C. P. Sloan, J. C. Cuevas, C. Quesnelle and V. Snieckus, *Tetrahedron Lett.* **29**, 4685 (1988).
80. K. A. Parker, D. M. Spero and K. C. Inman, *Tetrahedron Lett.* **27**, 2833 (1986).
81. M. D. Bachi and E. Bosch, *Heterocycles* **28**, 579 (1989); M. D. Bachi and D. Denemark, *Heterocycles* **28**, 583 (1989).
82. M. D. Bachi and E. Bosch, *Tetrahedron Lett.* **27**, 641 (1986).
83. D. L. Boger and R. Mathvink, *J. Org. Chem.* **53**, 3377 (1988).
84. A. G. Angoh and D. L. J. Clive, *J. Chem. Soc. Chem. Commun.* 980 (1985).
85. M. D. Bachi and E. Bosch, *J. Org. Chem.* **54**, 1234 (1989).
86. D. Batty and D. Crich, *Synthesis* 273 (1990).
87. D. L. J. Clive, P. L. Beaulieu and L. Set, *J. Org. Chem.* **49**, 1313 (1984).
88. R. Tsang and B. Fraser-Reid, *J. Am. Chem. Soc.* **108**, 8102 (1986).
89. A. L. J. Beckwith and B. P. Hay, *J. Am. Chem. Soc.* **111**, 230 (1989).
90. P. A. Bartlett, K. L. McLaren and P. C. Ting, *J. Am. Chem. Soc.* **110**, 1633 (1988).
91. G. Buchi and H. Wuest, *J. Org. Chem.* **44**, 546 (1979).
92. P. Bakuzis, O. O. S. Campos and M. L. F. Bakuzis, *J. Org. Chem.* **41**, 3261 (1976).
93. M. Ladlow and G. Pattenden, *J. Chem. Soc. Perkin Trans. 1*, 1107 (1988).
94. G. Stork and N. H. Baine, *Tetrahedron Lett.* **26**, 5927 (1985).
95. F. L. Harris and L. Weiler, *Tetrahedron Lett.* **28**, 2941 (1987).
96. M. D. Bachi, F. Frolow and C. Hoornaert, *J. Org. Chem.* **48**, 1841 (1983).
97. G. Just and G. Sacripante, *Can. J. Chem.* **65**, 104 (1987).
98. D. Batty, D. Crich and S. M. Fortt, *J. Chem. Soc. Chem. Commun.* 1366 (1989).
99. H. Finch, L. M. Harwood, G. M. Robertson and R̆. C. Sewell, *Tetrahedron Lett.* **30**, 2585 (1989).
100. T. Ghosh and H. Hart, *J. Org. Chem.* **53**, 2396 (1988).
101. S. A. Ahmad-Junan and D. A. Whiting, *J. Chem. Soc. Chem. Commun.* 1160 (1988).
102. S. A. Ahmad-Junan, A. J. Walkington and D. A. Whiting, *J. Chem. Soc. Chem. Commun.* 1613 (1989).

103. R. Tsang and B. Fraser-Reid, *J. Am. Chem. Soc.* **108**, 2116 (1986); R. Tsang, J. K. Dickson, H. Pak, R. Walton and B. Fraser-Reid, *J. Am. Chem. Soc.* **109**, 3484 (1987).
104. H. Riemann, A. S. Capomaggi, T. Strauss, E. P. Oliveto and D. H. R. Barton, *J. Am. Chem. Soc.* **83**, 4481 (1961).
105. M. Barbier, D. H. R. Barton, M. Devys and R. S. Topgi, *J. Chem. Soc. Chem. Commun.* 743 (1984).
106. P. Dowd and S. C. Choi, *Tetrahedron* **45**, 77 (1989).
107. A. L. J. Beckwith, R. Kazlauskas, M. R. SynerLyons, *J. Org. Chem.* **48**, 4718 (1983); D. E. O'Dell, J. T. Loper and T. L. Macdonald, *J. Org. Chem.* **53**, 5225 (1988).
108. J. E. Baldwin, R. M. Adlington and J. Robertson, *Tetrahedron* **45**, 909 (1989).
109. N. A. Porter, V. H. T. Chang, D. R. Magnin and B. T. Wright, *J. Am. Chem. Soc.* **110**, 3554 (1988); N. A. Porter and V. H. T. Chang, *J. Am. Chem. Soc.* **109**, 4976 (1987).
110. N. J. G. Cox, G. Pattenden and S. D. Mills, *Tetrahedron Lett.* **30**, 621 (1989).
111. N. A. Porter, D. R. Magnin and B. T. Wright, *J. Am. Chem. Soc.*, **108**, 2787 (1986).
112. A. J. Bloodworth, D. Crich and T. Melvin, *J. Chem. Soc. Chem. Commun.* 786 (1987).
113. E. D. Brown, T. W. Sam, J. K. Sutherland and A. Torre, *J. Chem. Soc. Perkin Trans. 1*, 2326 (1975).
114. L. A. Paquette, J. A. Colapret and D. R. Andrews, *J. Org. Chem.* **50**, 201 (1985).
115. L. M. van der Linde and A. J. A. van der Weerdt, *Tetrahedron Lett.* **25**, 1201 (1984).
116. L. Friedmann, *J. Am. Chem. Soc.* **86**, 1885 (1964).
117. I. De-Riggi, J. M. Surzur and M. P. Bertrand, *Tetrahedron* **44**, 7119 (1988).
118. D. C. Lathbury, P. J. Parsons and I. Pinto, *J. Chem. Soc. Chem. Commun.* 81 (1988).
119. D. P. Curran, D. Kim, H. T. Liu and W. Shen, *J. Am. Chem. Soc.* **110**, 5900 (1988); V. Snieckus, J. C. Cuevas, C. P. Sloan, H. Liu and D. P. Curran, *J. Am. Chem. Soc.* **112**, 896 (1990).
120. M. Julia, *Acc. Chem. Res.* (1971), **4**, 386; M. Julia, *Pure Appl. Chem.* **15**, 167 (1967).
121. M. Julia and C. James, *Compt. Rendu Ser. C.* **255**, 959 (1962).
122. M. Julia, J. M. Surzur, L. Katz and F. Le Goffic, *Bull. Soc. Chim. Fr.* 1116 (1964).
123. G. Stork and R. Mook, *J. Am. Chem. Soc.* **105**, 3720 (1983).
124. D. P. Curran and D. M. Rakiewicz, *Tetrahedron* **41**, 3943 (1985); D. P. Curran and S. C. Kuo, *J. Am. Chem. Soc.* **108**, 1106 (1986).
125. P. J. Parsons, P. A. Willis and S. C. Eyley, *J. Chem. Soc. Chem. Commun.* 283 (1988).
126. A. L. J. Beckwith, D. H. Roberts, C. H. Schiesser and A. Wallner, *Tetrahedron Lett.* **26**, 3349 (1985).
127. M. Julia, F. Le Goffic and L. Katz, *Bull. Soc. Chim. Fr.* 1122 (1964).
128. J. D. Winkler and V. Sridar, *J. Am. Chem. Soc.* **108**, 1708 (1986); J. D. Winkler and V. Sridar, *Tetrahedron Lett.* **29**, 6219 (1988).
129. D. H. R. Barton, E. da Silva and S. Z. Zard, *J. Chem. Soc. Chem. Commun.* 285 (1988).
130. G. Stork, P. M. Sher and H. L. Chen, *J. Am. Chem. Soc.* **108**, 6384 (1986); for a similar cyclization/alkylation see: H. Togo and O. Kikuchi, *Tetrahedron Lett.* **29**, 4133 (1988).
131. K. Miura, K. Fugami, K. Oshima, and K. Utimoto, *Tetrahedron Lett.* **29**, 1543 (1988).
132. J. D. Harling and W. B. Motherwell, *J. Chem. Soc. Chem. Commun.* 1380 (1988).
133. T. Miura, K. Fugami, K. Oshima and K. Utimoto, *Tetrahedron Lett.* **29**, 5135 (1988); K. S. Feldman, A. L. Romanelli, R. E. Ruckle and R. F. Miller, *J. Am. Chem. Soc.* **110**, 3300 (1988).
134. A. Bury, S. T. Corker and M. D. Johnson, *J. Chem. Soc. Perkin Trans. 1*, 645 (1982); M. R. Ashcroft, P. Bougeard, A. Bury, C. J. Cooksey, and M. D. Johnson, *J. Organomet. Chem.* **289**, 403 (1985); P. Bougeard, C. J. Cooksey, M. D. Johnson, M. J. Lewin, S. Mitchell and P. A. Owens, *J. Organomet. Chem.* **288**, 349 (1985); M. R. Ashcroft, A. Bury, C. J. Cooksey, A. G. Davies, B. D. Gupta, M. D. Johnson and H. Morris, *J. Organomet. Chem.* **195**, 89 (1980).

Index